More Precisely

More Precisely

The Math You Need to do Philosophy

Eric Steinhart

BROADVIEW GUIDES to PHILOSOPHY

Library and Archives Canada Cataloguing in Publication

Steinhart, Eric
More precisely : the math you need to do philosophy / Eric Steinhart.

(Broadview guides to philosophy)
Includes bibliographical references and index.
ISBN 978-1-55111-909-0

1. Logic, Symbolic and mathematical. 2. Mathematics—Philosophy.
3. Philosophy. I. Title. II. Series.
QA9.S745 2008 511.3 C2008-906489-5

Broadview Press is an independent, international publishing house, incorporated in 1985. Broadview believes in shared ownership, both with its employees and with the general public; since the year 2000 Broadview shares have traded publicly on the Toronto Venture Exchange under the symbol BDP. We welcome comments and suggestions regarding any aspect of our publications—please feel free to contact us at the addresses below or at broadview@broadviewpress.com.

North America
PO Box 1243,
Peterborough, Ontario,
Canada K9J 7H5

PO Box 1015, 2215 Kenmore Avenue,
Buffalo, NY, USA 14207
Tel: (705) 743-8990
Fax: (705) 743-8353
email: customerservice@broadviewpress.com

UK, Ireland, and continental Europe
NBN International, Estover Road,
Plymouth, UK PL6 7PY
Tel: 44 (0) 1752 202300
Fax: 44 (0) 1752 202330
email: enquiries@nbninternational.com

Australia and New Zealand
UNIREPS, University of New South Wales
Sydney, NSW, Australia 2052
Tel: 61 2 9664 0999
Fax: 61 2 9664 5420
email: infopress@unsw.edu.au

www.broadviewpress.com

This book is printed on paper containing 100% post-consumer fibre.

PROCESSED CHLORINE FREE
PROCÉDÉ SANS CHLORE

PRINTED IN CANADA

Contents

3. Machines

4. Semantics

5. Probability

6. Utilitarianism

7. From the Finite to the Infinite

Preface

Anyone doing philosophy today needs to have a sound understanding of a wide range of basic mathematical concepts. Unfortunately, most applied mathematics texts are designed to meet the needs of scientists. And much of the math used in the sciences is not used in philosophy. You're looking at a *mathematics book* that's designed to meet the needs of philosophers. *More Precisely* introduces the mathematical concepts you need in order to do philosophy today. As we introduce these concepts, we illustrate them with many classical and recent philosophical examples. This is math for *philosophers*.

It's important to understand what *More Precisely* is and what it isn't. *More Precisely* is *not* a philosophy of mathematics book. It's a mathematics book. We're not going to talk philosophically about mathematics. We are going to teach you the mathematics you need to do philosophy. We won't enter into any of the debates that rage in the current philosophy of math. Do abstract objects exist? How do we have mathematical knowledge? We don't deal with those issues here. *More Precisely* is *not* a logic book. We're not introducing you to any logical calculus. We won't prove any theorems. You can do a lot of math with very little logic. If you know some propositional or predicate logic, it won't hurt. But even if you've never heard of them, you'll do just fine here. *More Precisely* is an *introductory* book. It is not an advanced text. It aims to cover the basics so that you're prepared to go into the depths on the topics that are of special interest to you. To follow what we're doing here, you don't need anything beyond high school mathematics. We introduce all the technical notations and concepts gently and with many examples. We'll draw our examples from many branches of philosophy – including metaphysics, philosophy of mind, philosophy of language, epistemology, ethics, and philosophy of religion.

It's natural to start with some set theory. All branches of philosophy today make some use of set theory. If you want to be able to follow what's going on in philosophy today, you need to master at least the basic language of set theory. You need to understand the specialized notation and vocabulary used to talk about sets. For example, you need to understand the concept of the intersection of two sets, and to know how it is written in the specialized notation of set theory. Since we're not doing philosophy of math, we aren't going to get into any debates about whether or not sets exist. Before getting into such debates, you need to have a clear understanding of the objects you're arguing about. Our purpose in Chapter 1 is to introduce you to the language of sets and the basic ideas of set theory. Chapter 2 introduces relations and functions. Basic set-theoretic notions, especially relations and functions, are used extensively in the later chapters. So if you're not familiar with those notions, you've got to start with Chapters 1 and 2. Make sure you've really mastered the ideas in Chapters 1 and 2 before going on.

After we discuss basic set-theoretic concepts, we go into concepts that are used in various branches of philosophy. Chapter 3 introduces machines. A machine (in the special sense used in philosophy, computer science, and mathematics) isn't an industrial device. It's a formal structure used to describe some lawful pattern of activity. Machines are often used in philosophy of mind – many philosophers model minds as machines. Machines are sometimes used in metaphysics – simple universes can be modeled as networks of interacting machines. You can use these models to study space, time, and causality. Chapter 4 introduces some of the math used in the philosophy of language. Sets, relations, and functions are extensively used in formal semantic theories – especially possible worlds semantics. Chapter 5 introduces basic probability theory. Probability theory is used in epistemology and the philosophy of science (e.g., Bayesian epistemology, Bayesian confirmation theory). Mathematical concepts are often used in ethics. Utilitarians make great use of sums and products – the utility of a possible world is the sum of the happinesses of the persons in that world. So Chapter 6 discusses some of the math used in various utilitarian theories. Finally, the topic of infinity comes up in many philosophical discussions. Is the mind finitely or infinitely complex? Can infinitely many tasks be done in finite time? What does it mean to say that God is infinite? Chapter 7 introduces the notion of recursion and countable infinity. Chapter 8 shows that there is an endless progression of bigger and bigger infinities. It introduces transfinite recursion.

We illustrate the mathematical concepts with philosophical examples. We aren't interested in the philosophical soundness of these examples. As mentioned, we devote a chapter to the kinds of mathematics used in some utilitarian theories. Is this because utilitarianism is right? It doesn't matter. What matters is that utilitarianism uses lots of math, and that you need to know that math before you can really understand utilitarianism. As another example, we'll spend many pages explaining the mathematical apparatus behind various versions of possible worlds semantics. Is this because possible worlds really exist? We don't care. We do care that possible worlds semantics makes heavy use of sets, relations, and functions. As we develop the mathematics used in philosophy, we obviously talk about lots and lots of mathematical objects. We talk about sets, numbers, functions, and so on. Our attitude to these objects is entirely uncritical. We're engaged in exposition, not evaluation. We leave the interpretations and evaluations up to you. Although we aim to avoid philosophical controversies, *More Precisely* is not a miscellaneous assortment of mathematical tools and techniques. If you look closely, you'll see that the ideas unfold in an orderly and connected way. *More Precisely* is a conceptual narrative.

Our hope is that learning the mathematics we present in *More Precisely* will help you to do philosophy. You'll be better equipped to read technical philosophical articles. Articles and ideas that once might have seemed far too formal will become easy to understand. And you'll be able to apply these concepts in your own philosophical thinking and writing. Of course, some philosophers might object: why should philosophy use mathematics at all? Shouldn't philosophy avoid technicalities? We agree that technicality for its own sake ought to be avoided. As Ansel Adams once said, "There's nothing worse than a sharp image of a fuzzy concept." A bad idea doesn't get any better by expressing it in formal terms. Still, we think that philosophy has a lot to gain from becoming more mathematical. As science became more mathematical, it became more successful. Many deep and ancient problems were solved by making mathematical models of various parts and aspects of the universe. Is it naive to think that philosophy can make similar progress? Perhaps. But the introduction of formal methods into philosophy in the last century has led to enormous gains in clarity and conceptual power. Metaphysics, epistemology, ethics, philosophy of language, philosophy of science, and many other branches of philosophy, have made incredible advances by using the best available mathematical tools. Our hope is that this conceptual progress, slow and uncertain as it may be, will gain even greater strength.

Additional resources for *More Precisely* are available on the World Wide Web. These resources include extra examples as well as exercises. For more information, please visit

<http://broadviewpress.com/moreprecisely>

or

<http://www.ericsteinhart.com>.

Many thanks are due to the philosophers who helped with this project. Jim Moor deserves thanks for helping with the structure of the project. I especially appreciate the help of Chris Daly, Bob Martin, Tara Lowes, and Kathleen Wallace. They carefully read all or part of the manuscript and made very valuable suggestions. And I'm grateful to Gillman Payette for an extremely close reading. His suggestions made this a much better text! Finally, I'd like to thank Ryan Chynces and Alex Sager for being wonderful editors.

1

SETS

1. Collections of Things

As the 19th century was coming to a close, many people began to try to think precisely about collections. Among the first was the Russian-German mathematician Georg Cantor. Cantor introduced the idea of a *set*. For Cantor, a set is the collection of many things into a whole (1955: 85). It's not hard to find examples of sets: a crowd of people is a set of people; a herd of cows is a set of cows, a fleet of ships is a set of ships, and so on. The things that are collected together into a set are the *members* of the set. So if a library is a set of books, then the books in the library are the members of the library. Likewise, if a galaxy of stars is a set of stars, then the stars in the galaxy are the members of the galaxy.

As time went by, Cantor's early work on sets quickly became an elaborate theory. Set theory went through a turbulent childhood (see van Heijenoort, 1967; Hallett, 1988). But by the middle of the 20th century, set theory had become stable and mature. Set theory today is a sophisticated branch of mathematics. Set theorists have developed a rich and complex technical vocabulary – a network of special terms. And they have developed a rich and complex system of rules for the correct use of those terms. Our purpose in this chapter is to introduce you to the vocabulary and rules of set theory. Why study set theory? Because it is used extensively in current philosophy. You need to know it.

Our approach to sets is uncritical. We take the words of the set theorists at face value. If they say some sets exist, we believe them. Of course, as philosophers, we have to look critically at the ideas behind set theory. We need to ask many questions about the assumptions of the set theorists. But before you can criticize set theory, you need to understand it. We are concerned here only with the understanding. You may or may not think that numbers exist. But you still need to know how to do arithmetic. Likewise, you may or may not think that sets exist. But to succeed in contemporary philosophy, you need to know at least some elementary set theory. Our goal is to help you master the set theory you need to do philosophy. Our approach to sets is informal. We introduce the notions of set theory step by step, little by little. A more formal approach involves the detailed study of the *axioms* of set theory. The axioms of set theory are the precisely stated rules of set theory. Studying the axioms of set theory is advanced work. So we won't go into the axioms here. Our aim is to *introduce* set theory. We can introduce it informally. Most importantly, in the coming chapters, we'll show how ideas from set theory (and other parts of mathematics) are applied in various branches of philosophy.

We start with the things that go into sets. After all, we can't have collections of things if we don't have any things to collect. We start with things that aren't sets. An *individual* is any thing that isn't a set. Sometimes individuals are known as *urelemente* (this is a German word pronounced oor-ella-mentuh, meaning *primordial, basic, original elements*). Beyond saying that individuals are not sets, we place no restrictions on the individuals. The individuals that can go into sets can be names, concepts, physical things, numbers, monads, angels, propositions, possible worlds, or whatever you want to think or talk about. So long as they aren't sets. Sets can be inside sets, but then they aren't counted as *individuals*. Given some individuals, we can collect them together to make sets. Of course, at some point we'll have to abandon the idea that every set is a construction of things collected together by someone. For example, set theorists say that there exists a set whose members are all finite numbers. But no human person has ever gathered all the finite numbers together into a set. Still, for the moment, we'll use that kind of constructive talk freely.

2. Sets and Members

Sets have names. One way to refer to a set is to list its members between curly braces. Hence the name of the set consisting of Socrates, Plato, and Aristotle is {Socrates, Plato, Aristotle}. Notice that listing the members is different from listing the names of the members. So

{Socrates, Plato, Aristotle} is a set of *philosophers*; but

{"Socrates", "Plato", "Aristotle"} is a set of *names* of philosophers.

The membership relation is expressed by the symbol $\in$. So we symbolize the fact that Socrates is a member of {Socrates, Plato} like this:

Socrates $\in$ {Socrates, Plato}.

The negation of the membership relation is expressed by the symbol $\notin$. We therefore symbolize the fact that Aristotle is not a member of {Socrates, Plato} like this:

Aristotle $\notin$ {Socrates, Plato}.

And we said that individuals don't have members (in other words, no object is a member of any non-set). So Socrates is not a member of Plato. Write it like this:

Socrates $\notin$ Plato.

Identity. Two sets are identical if, and only if, they have the same members. The long phrase "if and only if" indicates logical equivalence. To say a set S is identical with a set T is equivalent to saying that S and T have the same members. That is, if S and T have the same members, then they're identical; and if S and T are identical, then they have the same members. The phrase "if and only if" is often abbreviated as "iff". It's not a spelling mistake! Thus

S = T if and only if S and T have the same members

is abbreviated as

S = T iff S and T have the same members.

More precisely, a set S is identical with a set T iff for every x, x is in S iff x is in T. One of our goals is to help you get familiar with the symbolism of set theory. So we can write the identity relation between sets in symbols like this:

S = T iff (for every x)(($x \in$ S) iff ($x \in$ T)).

You can easily see that {Socrates, Plato} = {Socrates, Plato}. When writing the name of a set, the order in which the members are listed makes no difference. For example,

{Socrates, Plato} = {Plato, Socrates}.

When writing the name of a set, mentioning a member many times makes no difference. You only need to mention each member once. For example,

{Plato, Plato, Plato} = {Plato, Plato} = {Plato};

{Socrates, Plato, Socrates} = {Socrates, Plato}.

When writing the name of a set, using different names for the same members makes no difference. As we all know, Superman is Clark Kent and Batman is Bruce Wayne. So

{Superman, Clark Kent} = {Clark Kent} = {Superman};

{Superman, Batman} = {Clark Kent, Bruce Wayne}.

Two sets are distinct if, and only if, they have distinct members:

{Socrates, Plato} is not identical with {Socrates, Aristotle}.

3. Set Builder Notation

So far we've defined sets by listing their members. We can also define a set by giving a formula that is true of every member of the set. For instance, consider the set of happy things. Every member in that set is happy. It is the set of all x such that x is happy. We use a special notation to describe this set:

the set of . . .	$\{ \ldots \}$
the set of all x . . .	$\{ x \ldots \}$
the set of all x such that . . .	$\{ x \mid \ldots \}$
the set of all x such that x is happy	$\{ x \mid x \text{ is happy} \}$.

Note that we use the vertical stroke "|" to mean "such that". And when we use the variable x by itself in the set builder notation, the scope of that variable is wide open – x can be *anything*. Many sets can be defined using this set-builder notation:

the books in the library = $\{ x \mid x \text{ is a book and } x \text{ is in the library} \}$;

the sons of rich men = $\{ x \mid x \text{ is the son of some rich man} \}$.

A set is never a member of itself. At least not in standard set theory. There are some non-standard theories that allow sets to be members of themselves (see Aczel, 1988). But we're developing standard set theory here. And since standard set theory is used to define the non-standard set theories, you need to start with it anyway! Any definition of a set that implies that it is a member of itself is ill-formed – it does not in fact define any set at all. For example, consider the formula

the set of all sets = $\{ x \mid x \text{ is a set} \}$.

Since the set of all sets is a set, it must be a member of itself. But we've ruled out such ill-formed collections. A set that is a member of itself is a kind of vicious circularity. The rules of set theory forbid the formation of any set that is a member of itself. Perhaps there is a *collection* of all sets. But such a collection can't be a *set*.

4. Subsets

Subset. Sets stand to one another in various relations. One of the most basic relations is the subset relation. A set S *is a subset of* a set T iff every member of S is in T. More precisely, a set S is a subset of a set T iff for every x, if x is in S, then x is in T. Hence

{Socrates, Plato} is a subset of {Socrates, Plato, Aristotle}.

Set theorists use a special symbol to indicate that S is a subset of T:

$S \subseteq T$ means S is a subset of T.

Hence

$\{\text{Socrates, Plato}\} \subseteq \{\text{Socrates, Plato, Aristotle}\}$.

We can use symbols to define the subset relation like this:

$S \subseteq T$ iff (for every x)(if $x \in S$ then $x \in T$).

Obviously, if x is in S, then x is in S; hence every set is a subset of itself. That is, for any set S, $S \subseteq S$. For example,

{Socrates, Plato} is a subset of {Socrates, Plato}.

But remember that no set is a member of itself. Being a subset of S is different from being a member of S. The fact that $S \subseteq S$ does *not* imply that $S \in S$.

Proper Subset. We often want to talk about the subsets of S that are distinct from S. A subset of S that is not S itself is a *proper* subset of S. An identical subset is an *improper* subset. So

{Socrates, Plato} is an improper subset of {Socrates, Plato};

while

{Socrates, Plato} is a proper subset of {Socrates, Plato, Aristotle}.

We use a special symbol to distinguish proper subsets:

$S \subset T$ means S is a proper subset of T.

Every proper subset is a subset. So if $S \subset T$, then $S \subseteq T$. However, not every subset is a proper subset. So if $S \subseteq T$, it does not follow that $S \subset T$. Consider:

$\{\text{Socrates, Plato}\} \subseteq \{\text{Socrates, Plato, Aristotle}\}$	True
$\{\text{Socrates, Plato}\} \subset \{\text{Socrates, Plato, Aristotle}\}$	True
$\{\text{Socrates, Plato, Aristotle}\} \subseteq \{\text{Socrates, Plato, Aristotle}\}$	True
$\{\text{Socrates, Plato, Aristotle}\} \subset \{\text{Socrates, Plato, Aristotle}\}$	False

Two sets are identical iff each is a subset of the other:

$S = T$ iff $((S \subseteq T)\ \&\ (T \subseteq S))$.

Superset. A superset is the converse of a subset. If S is a subset of T, then T is a superset of S. We write it like this: T $\supseteq$ S means T is a superset of S. For example,

{Socrates, Plato, Aristotle} $\supseteq$ {Socrates, Plato}.

5. Small Sets

Unit Sets. Some sets contain one and only one member. A *unit set* is a set that contains exactly one member. For instance, the unit set of Socrates contains Socrates and only Socrates. The unit set of Socrates is {Socrates}. For any thing x, the unit set of x is $\{x\}$. Sometimes unit sets are known as *singleton* sets. Thus {Socrates} is a singleton set.

Some philosophers have worried about the existence of unit sets (see Goodman, 1956; Lewis, 1991: sec 2.1). Nevertheless, from our uncritical point of view, these worries aren't our concern. Set theorists tell us that there are unit sets, so we accept their word uncritically. They tell us that for every x, there exists a unit set $\{x\}$. And they tell us that x is not identical to $\{x\}$. On the contrary, $\{x\}$ is distinct from x. For example, {Socrates} is distinct from Socrates. Socrates is a person; {Socrates} is a set. Consider the set of all x such that x is a philosopher who wrote the *Monadology*. This is written formally as:

{ x | x is a philosopher who wrote the *Monadology* }.

Assuming that Leibniz is the one and only philosopher who wrote the *Monadology*, then it follows that this set is {Leibniz}.

Empty Set. The set of all x such that x is a dog is { x | x is a dog }. No doubt this is a large set with many interesting members. But what about the set of all x such that x is a unicorn? Since there are no unicorns, this set does not have any members. Or at least the set of actual unicorns has no members. Likewise, the set of all actual elves has no members. So the set of all actual unicorns is identical to the set of all actual elves.

A set that does not contain any members is said to be *empty*. Since sets are identified by their members, there is exactly one empty set. More precisely,

S is the empty set iff (for every x)(x is not a member of S).

Two symbols are commonly used to refer to the empty set:

$\varnothing$ is the empty set; and
{} is the empty set.

We'll use "{}" to denote the empty set. For example,

{} = { x | x is an actual unicorn };
{} = { x | x is a married bachelor }.

It is important not to be confused about the empty set. The empty set isn't nothing or non-being. If you think the empty set exists, then obviously you can't think that it is nothing. That would be absurd. The empty set is exactly what the formalism says it is: it is a set that does not contain any thing as a member. It is a set with no members. According to set theory, the empty set is an existing particular object.

The empty set is a subset of every set. Consider an example: {} is a subset of {Plato, Socrates}. The idea is this: for any x, if x is in {}, then x is in {Plato, Socrates}. How can this be? Well, pick some object for x. Let x be Aristotle. Is Aristotle in {}? The answer is no. So the statement "Aristotle is in {}" is false. And obviously, "Aristotle is in {Plato, Socrates} is false. Aristotle is not in that set. But logic tells us that the only way an *if-then* statement can be false is when the *if* part is true and the *then* part is false. Thus (somewhat at odds with ordinary talk) logicians count an *if-then* statement with a false *if* part as true. So even though both the *if* part and the *then* part of the whole *if-then* statement are false, the whole *if-then* statement "if Aristotle is in {}, then Aristotle is in {Socrates, Plato}" is true. The same reasoning holds for any object you choose for x. Thus for any set S, and for any object x, the statement "if x is in {}, then x is in S" is true. Hence {} is a subset of S.

We can work this out more formally. For any set S, $\{\} \subseteq$ S. Here's the proof: for any x, it is not the case that ($x \in \{\}$). Recall that when the antecedent (the *if* part) of a conditional is false, the whole conditional is true. That is, for any Q, when P is false, (if P then Q) is true. So for any set S, and for any object x, it is true that (if $x \in \{\}$ then x is in S). So for any set S, it is true that (for all x)(if $x \in \{\}$ then x is in S). Hence for any set S, $\{\} \subseteq$ S.

Bear this clearly in mind: the fact that {} is a *subset* of every set does *not* imply that {} is a *member* of every set. The subset relation is *not* the membership relation. Every set has the empty set as a subset. But if we want the empty set to be a member of a set, we have to put it into the set. Thus {A} has the empty set as a subset while {{}, A} has the empty set as both a subset and as a member. Clearly, {A} is not identical to {{}, A}.

6. Unions of Sets

Unions. Given any two sets S and T, we can take their *union*. Informally, you get the union of two sets by adding them together. For instance, if the Greeks = {Socrates, Plato} and the Germans = {Kant, Hegel}, then the union of the

Greeks and the Germans is {Socrates, Plato, Kant, Hegel}. We use a special symbol to indicate unions:

the union of S and T = S union T = $S \cup T$.

For example,

{Socrates, Plato} $\cup$ {Kant, Hegel} = {Socrates, Plato, Kant, Hegel}.

When forming the union of two sets, any common members are only included once:

{Socrates, Plato} $\cup$ {Plato, Aristotle} = {Socrates, Plato, Aristotle}.

The union of a set with itself is just that very same set:

{Socrates} $\cup$ {Socrates} = {Socrates}.

More formally, the union of S and T is the set that contains every object that is either in S or in T. Thus x is in the union of S and T iff x is in S only, or x is in T only, or x is in both. The union of S and T is the set of all x such that x is in S or x is in T. In symbols,

S union T = { x | x is in S or x is in T };

$S \cup T = \{ x \mid x \in S \text{ or } x \in T \}$.

Just as you can add many numbers together, so you can union many sets together. Just as you can calculate 2+3+6, so you can calculate $S \cup T \cup Z$. For example, {Socrates, Plato} $\cup$ {Kant, Hegel} $\cup$ {Plotinus} is {Socrates, Plato, Kant, Hegel, Plotinus}.

When a set is unioned with the empty set, the result is itself: $S \cup \{\} = S$. The union operator is defined only for *sets*. So the union of an individual with a set is undefined as is the union of two individuals.

7. Intersections of Sets

Intersections. Given any two sets S and T, we can take their *intersection*. Informally, you get the intersection of two sets by taking what they have in common. For instance, if the Philosophers = {Socrates, Aristotle} and the Macedonians = {Aristotle, Alexander}, then the intersection of the Philosophers with the Macedonians = {Aristotle}. We use a special symbol to indicate intersections:

the intersection of S and T = S intersection T = $S \cap T$.

For example,

{Socrates, Aristotle} $\cap$ {Aristotle, Alexander} = {Aristotle}.

The intersection of a set with itself is just that very same set:

{Socrates} $\cap$ {Socrates} = {Socrates}.

More formally, the intersection of S and T is the set that contains every object that is both in S and in T. Thus *x* is in the intersection of S and T iff both *x* is in S and *x* is in T. The intersection of S and T is the set of all *x* such that *x* is in S and *x* is in T. Thus

S intersection T = { *x* | *x* is in S and *x* is in T }.

Of course, we can use the ampersand "&" to symbolize "and". So we can write the intersection of S and T entirely in symbols as

$$S \cap T = \{ x \mid x \in S \,\&\, x \in T \}.$$

Disjoint Sets. Sometimes two sets have no members in common. Such sets are said to be disjoint. Set S is disjoint from set T iff there is no *x* such that *x* is in S and *x* is in T. Remember that the intersection of S and T is the *set* of all members that S and T have in common. So if S and T have no members in common, then their intersection is a set that contains no members. It is the empty set.

If two sets are disjoint, then their intersection contains no members. For example, the intersection of {Socrates, Plato} and {Kant, Hegel} contains no members. The two sets of philosophers are disjoint. There is no philosopher who is a member of both sets. Since these sets are disjoint, their intersection is the empty set.

The fact that two sets are disjoint does *not* imply that the empty set is a member of those sets. You can see that {A} $\cap$ {B} = {}. This does *not* imply that {} is a member of either {A} or {B}. If we wanted to define a set that contains both {} and A, we would write {{}, A}. Clearly {{}, A} is distinct from {A} and {{}, B} is distinct from {B}. And equally clearly, {{}, A} and {{}, B} are not disjoint. The empty set is a member of both of those sets. Consequently, their intersection is {{}}.

Since the empty set contains no members, it is disjoint from every set. The intersection of any set with the empty set is empty: S $\cap$ {} = {}. The intersection operator (like the union operator) is defined only for *sets*. So the intersection of an individual with a set is undefined as is the intersection of two individuals.

8. Difference of Sets

Given any two sets S and T, we can take their *difference*. Informally, you get the difference between two sets S and T by listing what is in S and not in T. For instance, if the Philosophers = {Socrates, Aristotle, Seneca} and the Greeks = {Socrates, Aristotle}, then the difference between the Philosophers and the Greeks = {Seneca}. Clearly, the difference between S and T is like subtracting T from S. We use a special symbol to indicate differences:

the difference between S and T = (S – T).

Hence

{Socrates, Aristotle, Seneca} – {Socrates, Aristotle} = {Seneca}.

Difference. More formally, the difference between S and T is the set of all x such that x is in S and x is not in T. In symbols,

$$S - T = \{ x \mid x \text{ is in S and } x \text{ is not in T} \};$$

$$S - T = \{ x \mid x \in S \;\&\; x \notin T \}.$$

The difference between any set and itself is empty: S – S = {}.

9. Set Algebra

The set operations of union, intersection, and difference can be combined. And sometimes different combinations define the same set. The different combinations are equal. The rules for these equalities are the algebra of set operations. For example, for any sets S and T and Z, we have these equalities:

$$S \cap (T \cup Z) = (S \cap T) \cup (S \cap Z);$$

$$S \cup (T \cap Z) = (S \cup T) \cap (S \cup Z);$$

$$S \cap (T - Z) = (S \cap T) - (S \cap Z);$$

$$(S \cap T) - Z = (S - Z) \cap (T - Z).$$

We won't prove these equalities here. You can find proofs in more technical books on set theory. The algebra of sets doesn't play a large role in philosophical work; but you should be aware that there is such an algebra.

10. Sets of Sets

So far, we've mostly talked about sets of individuals. But we can also form sets of sets. Given some sets, you can unify them into a set of sets. For example, suppose A, B, C, and D are some individuals. We can form the sets {A, B} and {C, D}. Now we can gather them together to make the set {{A, B}, {C, D}}. Making a set out of sets is not the same as taking the union of the sets. For set theory,

{{A, B}, {C, D}} is not identical to {A, B, C, D}.

When we work with sets of sets, we have to be very careful about the distinction between members and subsets. Consider the sets {A} and {A, {A}}. The individual A is a member of {A} and it is a member of {A, {A}}. The set {A} is not a member of {A}, but it is a subset of {A}. The set {A} is a member of {A, {A}}, and it is a subset of {A, {A}}. Consider the sets {A} and {{A}}. The individual A is a member of {A}; the set {A} is a member of the set {{A}}. But A is not a member of {{A}}. Hence the set {A} is not a subset of the set {{A}}. The member of {A} is not a member of {{A}}.

Ranks. Since we can make sets of sets, we can organize sets into *levels* or *ranks*. They are levels of membership – that is, they are levels of complexity with respect to the membership relation. The zeroth rank is the level of individuals (non-sets). Naturally, if we want to think of this rank as a single thing, we have to think of it as a set. For example, if A, B, and C are the only individuals, then

rank 0 = {A, B, C}.

The first rank is the rank of sets of individuals. Sets of individuals go on rank 1. The first rank is where sets first appear. Since the empty set is a set (it's a set of no individuals), it goes on the first rank. Here's the first rank:

rank 1 = {{}, {A}, {B}, {C}, {A, B}, {A, C}, {B, C}, {A, B, C}}.

We say a set is on the second rank if, and only if, it contains a set from the first rank. Since {{A}} contains a set from the first rank, it is on the second rank. Likewise {{A}, {B}}. But what about {{A}, A}? Since {{A}, A} contains {A}, it is on the second rank. We can't list all the sets on the second rank in our example – writing them all down would take a lot of ink. We only list a few:

rank 2 = {{{}}, {{A}}, {{C}}, {{A}, A}, {{A}, B}, {{A, B}, {C}}, . . . }.

As a rule, we say that a set is on rank n+1 if, and only if, it contains a set of rank n as a member. Set theory recognizes an endless hierarchy of ranks. There are *individuals* on rank 0; then *sets* on rank 1, rank 2, rank 3, . . . rank n, rank n+1, and so on without end. Figure 1.1 shows some individuals on rank 0. Above them, there are some sets on ranks 1 through 3. Not all the sets on those ranks are shown. The arrows are the membership relations. Each rank is to be thought of as a set, even though we don't write braces around the rank. We don't want to clutter our diagram too much.

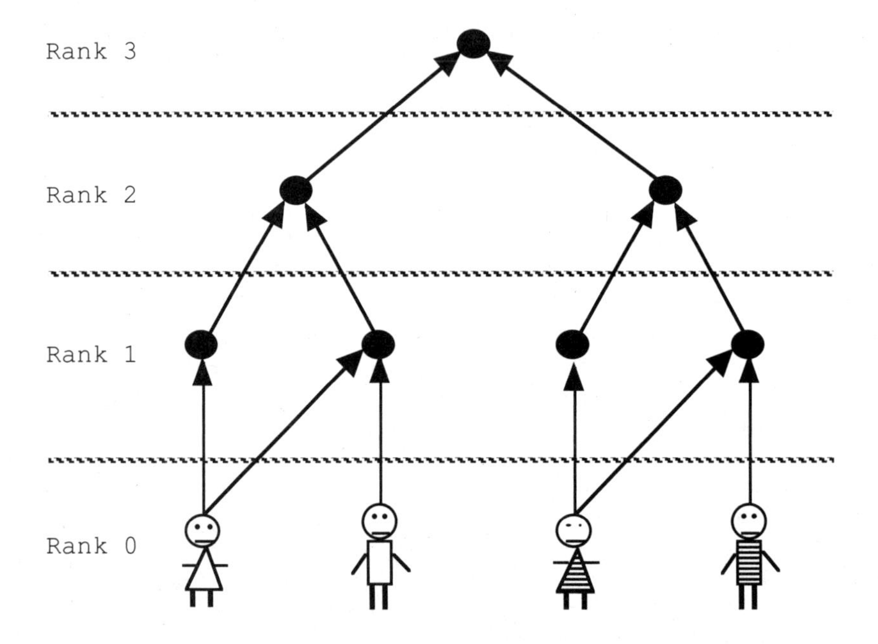

Figure 1.1 Some individuals and some sets on various ranks.

11. Union of a Set of Sets

Since we can apply the union operator to many sets, we can apply it to a set of sets. For example, if S and T and Z are all sets, then

$$\cup\{S, T, Z\} = S \cup T \cup Z.$$

Suppose S is a set of sets. The set S has members; since the members of S are sets, they too have members. Thus the set S has members of members. The union of S is the set of all the members of the members of S. For example,

$$\cup\{\{0\}, \{1, 2\}, \{3, 4\}\} = \{0, 1, 2, 3, 4\}.$$

If S is a set of sets, then $\cup S$ is the union of S. We can write it like this:

$$\text{the union of the members of } S = \bigcup_{x \in S} x.$$

An example will help make this notation clearer. Suppose some university has many libraries. We let L be the set of libraries:

L = { MathLibrary, ScienceLibrary, BusinessLibrary, MainLibrary }.

Each library is a set of books. If we want to obtain the set of all books held by the university, we need to take the union of the libraries. We write it like this:

$$\text{allTheLibraryBooks} = \bigcup_{x \in L} x.$$

Analogous remarks hold for intersections.

12. Power Sets

Power Set. Any set has a power set. The power set of a set S is the set of all ways of selecting members from S. Any way of selecting members from S is a subset of S. Hence the power set of S is the set of all subsets of S. That is

the power set of S = { x | x is a subset of S}.

Consider the set {A, B}. What are the subsets of {A, B}? The empty set {} is a subset of {A, B}. The set {A} is a subset of {A, B}. The set {B} is a subset of {A, B}. And of course {A, B} is a subset of itself. The power set of {A, B} contains these four subsets. Thus we say

the power set of {A, B} = {{}, {A}, {B}, {A, B}}.

The power set of S is sometimes symbolized by using a script version of the letter "P". Thus the power set of S is $\mathcal{P}$S. However, this can lead to confusion, since "P" is also used in discussions of probability. Another way to symbolize the power set of S is to write it as *pow S*. We'll use this notation. So in symbols,

pow S = { x | $x \subseteq$ S}.

Observe that *x is a member of* the power set of S iff *x is a subset of* S itself:

($x \in$ pow S) iff ($x \subseteq$ S).

Since the empty set {} is a subset of every set, the empty set is a member of the power set of S. Since S is a subset of itself, S is a member of the power set of S. And every other subset of S is a member of the power set of S. For example,

pow {0, 1, 2} = {{}, {0}, {1}, {2}, {0, 1}, {0, 2}, {1, 2}, {0, 1, 2}}.

The union of the power set of S is S. That is, $\cup$ (pow S) = S. For example,

$$\cup (\text{pow} \{A, B\}) = \cup \{\{\}, \{A\}, \{B\}, \{A, B\}\}$$

$$= \{\} \cup \{A\} \cup \{B\} \cup \{A, B\}$$

$$= \{A, B\}.$$

Again, the power set of S is the collection of all ways of selecting members from S. We can use a table to define a power set. Label the columns with the members of S. The rows are the subsets of S. A column is in a row (a member of S is in a subset of S) if there is a 1 in that column at that row. A column is not in a row (a member of S is not in a subset of S) if there is a 0 in that column in that row. Table 1.1 shows the formation of the power set of {Socrates, Plato, Aristotle}. If you are familiar with basic symbolic logic, you'll have recognized that the use of 1 and 0 clearly parallels the use of True and False in making a truth-table. So there is an intimate parallel between the ways of forming subsets and the ways of assigning truth-values to propositions.

Socrates	Plato	Aristotle	
1	1	1	{Socrates, Plato, Aristotle}
1	1	0	{Socrates, Plato}
1	0	1	{Socrates, Aristotle}
1	0	0	{Socrates}
0	1	1	{Plato, Aristotle}
0	1	0	{Plato}
0	0	1	{Aristotle}
0	0	0	{}

Table 1.1 The power set of {Socrates, Plato, Aristotle}.

Whenever you need to consider all the possible combinations of a set of objects, you're using the notion of the power set. Here's an example. One reasonable way to think about space (or space-time) is to say that space is made of geometrical points. And so a reasonable way to think about regions in space (or space-time) is to identify them with sets of points. Every possible combination of points is a region. If P is the set of points in the universe, then pow P is the set of regions. One of the regions in pow P is the empty region or the null region. It's just the empty set. On the one hand, someone might argue that there is no such place as the empty region. It's no place at all – hence not a region. On this view, the set of regions is pow P minus the empty set. On the other hand, someone else might argue that the empty set does exist: it's one of the parts one might divide P into, even though there's nothing in this part. On this view, the set of regions is pow P. What's your assessment of this debate?

13. Sets and Selections

Although sets were initially presented as collections, we can also think of them as *selections*. Given some individuals, you make sets by selecting. We can drop the constructive notion that people make these selections by just thinking about ways of selecting or just selections. The ways of selecting exist whether or not people ever make the corresponding selections. For example, given the two individuals A and B, there are four selections: neither A nor B; A but not B; B but not A; and both A and B.

We can use a truth-table format to specify each selection. Table 1.2 shows the four selections from A and B. For example, the selection neither A nor B has a 0 under A and a 0 under B. The 0 means that it is false that A is selected and it is false that B is selected. The four ways of assigning truth-values to the objects A and B are the four sets you can form given A and B. Note that if there are n objects, then there are 2^n ways to select those objects. This is familiar from logic: if there are n propositions, there are 2^n ways to assign truth-values to those propositions.

A	B	
1	1	{A, B}
1	0	{A}
0	1	{B}
0	0	{}

Table 1.2 Selections over individuals A and B.

We can iterate the notion of selection. We do this by taking all the objects we've considered so far as the inputs to the next selection table. So far, we have six objects: A, B, {}, {B}, {A}, {A,B}. These are the heads of the columns in our next selection table. Once again we assign truth-values (really, these are selection values, with 1 = selected and 0 = not selected). Since there are 6 inputs, there are $2^6 = 64$ outputs. That is, there are 64 ways of selecting from 6 things. We can't write a table with all these selections – it would take up too much space. So we only include a few selections in Table 1.3. They are there just for the sake of illustration. There is some redundancy in our formation of selections. For instance, Table 1.3 includes the empty set, which we already formed in our first level selections. Likewise, Table 1.3 includes {A, B}. The redundancy is eliminated by just ignoring any selection that exists on some lower level. As we consider more complex selections on higher levels, we do not include any selection more than once.

A	B	{}	{B}	{A}	{A, B}	
1	1	1	1	1	1	{A, B, {}, {A}, {B}, {A,B}}
1	1	1	1	1	0	{A, B, {}, {A}, {B}}
1	1	0	0	0	0	{A, B}
1	0	0	0	1	0	{A, {A}}
0	0	1	1	0	0	{{}, {B}}
0	0	1	0	0	0	{{}}
0	0	0	0	1	0	{{A}}
0	0	0	0	0	0	{}

Table 1.3 Some selections on the second level.

We can, of course, continue to iterate the process of selection. We can form a third level selection table. Then a fourth level, and so on, endlessly. Of course, a set theorist doesn't really care about these tables. They are just ways to illustrate the levels of sets. For the set theorist, these levels exist whether or not we write down any tables. This way of defining sets was pioneered by von Neumann (1923).

14. Pure Sets

One way to build up a universe of sets goes like this: We start with some individuals on rank 0. Given these individuals, we build sets on rank 1. Then we build sets on rank 2. And we continue like this, building up our universe of sets rank by rank. This way of building a universe of sets is clearly useful. But the price of utility is uncertainty. What are the individuals on rank 0? Should we include only simple material particles? Should we also include space-time points? Should we consider possible individuals, or just actual individuals? To avoid any uncertainties, we can ignore individuals altogether.

Pure Sets. When we ignore individuals, we get pure sets. The empty set is the simplest pure set. It does not contain any individuals. Any set built out of the empty set is also pure. For example, given the empty set, we can build {{}}. This set is pure. Given {} and {{}}, we can build {{}, {{}}}. And we can also build {{{}}}. Both of these sets are pure. We can go on in this way, building more and more complicated pure sets out of simpler pure sets. We can generate an entire universe of pure sets. Of course, all this constructive talk is just metaphorical. We can abandon this constructive talk in favor of non-constructive talk. We can say that for any two pure sets x and y, there exists the pure set $\{x, y\}$. Or, for every pure set x, there exists the power set of x. This power set is also pure.

Why study pure sets? Consider an analogy with numbers. Sometimes, we use numbers to count specific collections. For instance, we might say that if little Timmy has 5 apples and 4 oranges, then he has 9 pieces of fruit. And if little Susie has 5 dolls and 4 building blocks, then she has 9 toys. Given cases like this, you can see a general pattern: 5 plus 4 is 9. You can think about numbers in a general way without thinking about any particular things. These are pure numbers. And thus you can form general laws about numbers themselves – e.g., the sum of any two even numbers is even; $x + y = y + x$. Studying pure sets is like studying pure numbers. When we study pure sets, we study the sets themselves. Hence we can study the relations among sets without any distractions. For example, we can study the membership relation in its pure form. We can form laws about the sets themselves. And pure sets will turn out to have some surprising uses.

Set theorists use the symbol "V" to denote the whole universe of pure sets. To obtain V, we work from the bottom up. We define a series of *partial universes.* The initial partial universe is V_0. The initial partial universe does not contain any individuals. Hence V_0 is just the empty set {}. The next partial universe is V_1. The partial universe V_1 is bigger than V_0. When we make the step from V_0 to V_1, we want to include every possible set that we can form from the members of V_0. We don't want to miss any sets. For any set x, the power set of x is the set of all possible sets that can be formed from the members of x. Hence V_1 is the power set of V_0. Since the only subset of {} is {} itself, the set of all subsets of {} is the set that contains {}. It is {{}}. Therefore, V_1 = {{}}. The next partial universe is V_2. This universe is maximally larger than V_1. Thus V_2 is the power set of V_1. Accordingly, V_2 = {{}, {{}}}.

A good way to picture the sequence of partial universes is to use selection tables. Each selection table has the sets in universe V_n as its columns. The sets in V_{n+1} are the rows. Since V_0 contains no sets, the selection table that moves from V_0 to V_1 does not contain any columns. There's no point in displaying a table with no columns. Consequently, our first selection table is the one that moves from V_1 to V_2. This selection table is shown in Table 1.4. The partial universe V_2 is the set that includes all and only the sets that appear on the rows in the rightmost column. Hence V_2 = {{}, {{}}}. There are two objects in V_2. Hence there are 4 ways to make selections over these objects. Table 1.5 shows the move from V_2 to V_3. Table 1.6 shows the move from V_3 to V_4. For every partial universe, there is a greater partial universe. For every V_n, there is V_{n+1}. And V_{n+1} is always the power set of V_n. But it would take an enormous amount of space to draw the table that moves from V_4 to V_5. Since V_4 contains 16 sets, V_5 contains 2^{16} sets; it contains 65,536 sets. We can't draw V_5. We're energetic, but not that energetic. You can see why set theorists use V to denote the universe of pure sets. It starts at a bottom point – the empty set – and then expands as it grows upwards.

We generate bigger and bigger partial universes by repeating or *iterating* this process endlessly. The result is a series of partial universes. Each new partial universe contains sets on higher ranks. Hence the series is referred to as a *hierarchy*. And since the partial universes are generated by repetition, it is an iterative hierarchy – *the iterative hierarchy of pure sets*. You can learn more about the iterative hierarchy of pure sets in Boolos (1971) and Wang (1974). Table 1.7 illustrates the first five ranks of the hierarchy.

Given	{}	
Selections	0	{}
	1	{{}}

Table 1.4 Moving from V_1 to V_2.

Given	{}	{{}}	
Selections	0	0	{}
	1	0	{{}}
	0	1	{{{}}}
	1	1	{{}, {{}}}

Table 1.5 Moving from V_2 to V_3.

Given	{}	{{}}	{{{}}}	{{}, {{}}}	
Selections	0	0	0	0	{}
	1	0	0	0	{{}}
	0	1	0	0	{{{}}}
	1	1	0	0	{{}, {{}}}
	0	0	1	0	{{{{}}}}
	1	0	1	0	{{}, {{{}}}}
	0	1	1	0	{{{}}, {{{}}}}
	1	1	1	0	{{}, {{}}, {{{}}}}
	0	0	0	1	{{{}, {{}}}}
	1	0	0	1	{{}, {{}, {{}}}}
	0	1	0	1	{{{}}, {{}, {{}}}}
	1	1	0	1	{{}, {{}}, {{}, {{}}}}
	0	0	1	1	{{{{}}}, {{}, {{}}}}
	1	0	1	1	{{}, {{{}}}, {{}, {{}}}}
	0	1	1	1	{{{}}, {{{}}}, {{}, {{}}}}
	1	1	1	1	{{}, {{}}, {{{}}}, {{}, {{}}}}

Table 1.6 Moving from V_3 to V_4.

Universe V_0 = the empty set {};
V_0 = {};

Universe V_1 = the power set of V_0;
V_1 = {{}};

Universe V_2 = the power set of V_1;
V_2 = {{}, {{}}};

Universe V_3 = the power set of V_2;
V_3 = {{}, {{}}, {{{}}}, {{} {{}}}};

Universe V_4 = the power set of V_3;
V_4 = {{}, {{}}, {{{}}}, {{} {{}}}, {{{{}}}}, {{{} {{}}}}, {{} {{{}}}},
{{} {{} {{}}}}, {{{}} {{{}}}}, {{{}} {{} {{}}}}, {{{{}}} {{} {{}}}},
{{} {{}} {{{}}}}}, {{} {{}} {{} {{}}}}, {{} {{{}}} {{} {{}}}},
{{{}} {{{}}} {{} {{}}}}, {{} {{}} {{{}}} {{} {{}}}}}.

Table 1.7 The first five partial universes of pure sets.

15. Sets and Numbers

The *ontology* of a science is the list of the sorts of things that the science talks about. What sorts of things are we talking about when we do pure mathematics? Pure sets have played a philosophically interesting role in the ontology of pure mathematics. Some writers have argued that all purely mathematical objects can be reduced to or identified with pure sets. For example, von Neumann showed how to identify the natural numbers with pure sets. According to von Neumann, 0 is the empty set {} and each next number $n+1$ is the set of all lesser numbers. That is, $n+1 = \{0, \ldots n\}$. Hence the von Neumann Way looks like this:

Von Neumann Way

0 = {};
1 = {0} = {{}};
2 = {0, 1} = {{}, {{}}};
3 = {0, 1, 2} = {{}, {{}}, {{}, {{}}}}.

Similar techniques let mathematicians identify all the objects of pure mathematics with pure sets. So it would seem that the whole universe of mathematical objects – numbers, functions, vectors, and so on – is in fact just the universe of pure sets V.

It would be nice to have an ontology that posits exactly one kind of object – an ontology of pure sets. However, the philosopher Paul Benacerraf raised an objection to this reduction of mathematical objects to pure sets. He focused on the reduction of numbers to pure sets. He argued in 1965 that there are many equally good ways to reduce numbers to pure sets. We've already mentioned the von Neumann way. But Benacerraf pointed out that the mathematician Zermelo gave another way. According to the Zermelo way, 0 is the empty set {} and each next number $n+1$ is the set of its previous number n. That is, $n+1 = \{n\}$. So the Zermelo way looks like this:

Zermelo Way

$0 = \{\}$;
$1 = \{0\} = \{\{\}\}$;
$2 = \{1\} = \{\{\{\}\}\}$;
$3 = \{2\} = \{\{\{\{\}\}\}\}$.

Apparently, there are (at least) two ways to reduce numbers to sets. So which is the right way? We can't say that 2 is both {{{}}} and {{}, {{}}}. That would be a contradiction. Since there is no single way to reduce 2 to a pure set, there really isn't any way to reduce 2 to a pure set. The same goes for all the other numbers. The multiplicity of reductions invalidates the very idea of reduction. And thus Benacerraf objects to the reduction of numbers to pure sets. Philosophers have responded to Benacerraf in many ways. One way is the movement known as *structuralism* in philosophy of mathematics (see Resnik, 1997; Shapiro, 1997). Another way is to deny that both reductions are equally good. It can be argued that the von Neumann way has many advantages (Steinhart, 2003). The issue is not settled. Some philosophers say that numbers (and all other mathematical objects) are not reducible to pure sets; others say that they are reducible to pure sets.

A more extreme view is that *all* things are reducible to pure sets. The idea is this: physical things are reducible to mathematical things; mathematical things are reducible to pure sets. Quine developed this view. He argued in several places that all things – whether mathematical or physical – are reducible to pure sets and therefore *are* pure sets (see Quine, 1976, 1978, 1981, 1986). To put it very roughly, material things can be reduced to the space-time regions they occupy; space-time regions can be reduced to sets of points; points can be reduced to their numerical coordinates; and, finally, numbers can be reduced to pure sets. Quine's ontology is clear: all you need is sets. This is a radical idea. Of course, we're not interested in whether it is true or not. We mention it only so that you can see that pure sets, despite their abstractness, can play an important role in philosophical thought. Still, the question can be raised: what do you think of Quine's proposal?

16. Sums of Sets of Numbers

When given some set of numbers, we can add them all together. For example, consider the set {1, 2, 3}. The sum of the numbers in this set is $1 + 2 + 3 = 6$.

The Greek letter Σ (*sigma*) is conventionally used to express the sum of a set of numbers. There are many ways to use Σ to express the sum of a set of numbers. One way is:

$$\Sigma\{1, 2, 3\} = 6.$$

Given a set of numbers A, another notation for expressing its sum looks like this:

the sum . . . $= \sum$

the sum, for all x in A, . . . $= \sum_{x \in A}$

the sum, for all x in A, of $x = \sum_{x \in A} x.$

17. Ordered Pairs

Sets are not ordered. The set {A, B} is identical with the set {B, A}. But suppose we want to indicate that A comes before B in some way. We indicate that A comes before B by writing the ordered pair (A, B). The ordered pair (A, B) is not identical with the ordered pair (B, A). When we write ordered pairs, the *order* makes a difference.

Since sets are not ordered, it might seem like we can't use sets to define ordered pairs. But we can. Of course, the ordered pair (A, B) is not identical with the set {A, B}. It is identical with some more complex set. We want to define a set in which A and B are ordered. We want to define a set that indicates that A comes first and B comes second. We can do this by indicating the two things in the ordered pair and by distinguishing the one that comes first. One way to do this is to distinguish the first thing from the second thing by picking it out and listing it separately alongside the two things. For example: in the set {{A}, {A, B}}, the thing A is listed separately from the two things {A, B}. We can thus identify the ordered pair (A, B) with the set {{A}, {A, B}}. Notice that {{A}, {A, B}} is *not* identical with {{B}, {A, B}}. Hence (A, B) is not identical with (B, A).

Ordered Pairs. An ordered pair of things is written (x, y). As a rule, for any two things x and y, the ordered pair (x, y) is the set $\{\{x\}, \{x, y\}\}$. Notice that (x, y) is not the same as (y, x), since $\{\{x\}, \{x, y\}\}$ is not the same as $\{\{y\}, \{x, y\}\}$.

Figure 1.2 shows the membership network for the ordered pair (x, y). Notice that the membership of x in $\{x, y\}$, but the failure of membership of y in $\{x\}$, introduces an asymmetry. This asymmetry is the source of the order of the set $\{\{x\}, \{x, y\}\}$.

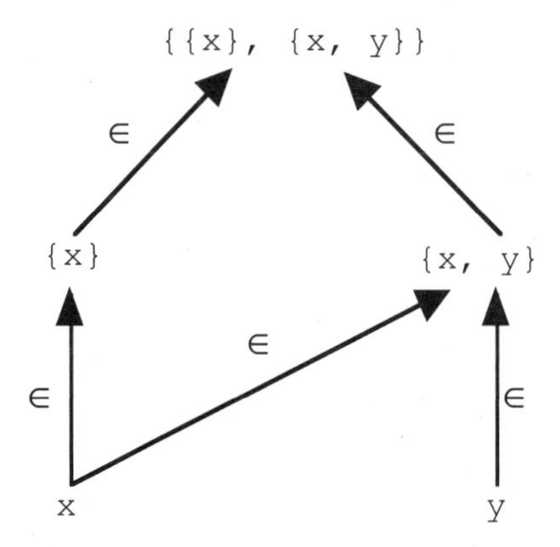

Figure 1.2 The ordered pair (x, y).

Notice that $(x, x) = \{\{x\}, \{x, x\}\}$. Recall that listing the same member many times makes no difference. Hence $\{x, x\} = \{x\}$. Thus we have $(x, x) = \{\{x\}, \{x, x\}\} = \{\{x\}, \{x\}\} = \{\{x\}\}$. Notice further that $\{\{x\}\}$ is not identical with either $\{x\}$ or x. The set $\{\{x\}\}$ exists two membership levels above x. All ordered pairs, even pairs whose elements are identical, exist two levels above their elements.

It is easy but tedious to prove that $(x, y) = (a, b)$ if, and only if, $x = a$ and $y = b$. This shows that order is essential for the identity of ordered pairs.

18. Ordered Tuples

Although ordered pairs are nice, we sometimes want to define longer orderings. An ordered triple is (x, y, z); an ordered quadruple is (w, x, y, z). And, more generally, an *ordered n-tuple* is $(x_1, \ldots x_n)$. Thus an ordered pair is an ordered 2-tuple; an ordered triple is an ordered 3-tuple. We can refer to an ordered n-tuple as just an n-tuple. And when n is clear from the context, or we don't care about it, we just refer to *tuples*.

Ordered Tuples. We define longer orderings by nesting ordered pairs. So the triple (x, y, z) is really the pair $((x, y), z)$. Notice that $((x, y), z)$ is an ordered pair whose first element is the ordered pair (x, y) and whose second element is z. So it is an ordered pair nested inside an ordered pair. The ordered quadruple (w, x, y, z) is $(((w, x), y), z)$. More generally, we define an $(n+1)$-tuple in terms of an n-tuple like this: $(x_1, \ldots x_{n+1}) = ((x_1, \ldots x_n), x_{n+1})$.

Notice that we can resolve an ordered 3-tuple into sets:

$$
\begin{aligned}
(x, y, z) &= ((x, y), z) \\
&= \{\{(x, y)\}, \{(x, y), z\}\} \\
&= \{\{\{\{x\}, \{x, y\}\}\}, \{\{\{x\}, \{x, y\}\}, z\}\}.
\end{aligned}
$$

We can repeat this process and resolve any n-tuple into sets. This shows that even short ordered n-tuples (such as ordered 3-tuples) contain lots of structure.

19. Cartesian Products

If we have two sets, we may want to form the set of all ordered pairs of elements from those two sets. For instance: suppose M is some set of males and F is some set of females; the set of all male-female ordered pairs is the set of all (x, y) such that x is in M and y is in F. Notice that this is distinct from the set of all female-male ordered pairs. The set of all female-male ordered pairs is the set of all (x, y) such that x is in F and y is in M.

There are indeed times when order does matter. The fact that Eric and Kathleen are a couple can be indicated by the set {Eric, Kathleen}. The fact that Eric is the husband of Kathleen can be indicated by the ordered pair (Eric, Kathleen), while the fact that Kathleen is the wife of Eric can be indicated by the ordered pair (Kathleen, Eric). So there is a difference between the sets {Eric, Kathleen} and the pairs (Kathleen, Eric) and (Eric, Kathleen).

Suppose we have a set of males M and a set of females F. The set of males is {A, B, C, D}, while the set of females is {1, 2, 3, 4}. We can use a table to pair off males with females. Table 1.8 shows how we form all the partnerships.

	1	2	3	4
A	(A,1)	(A,2)	(A,3)	(A,4)
B	(B,1)	(B,2)	(B,3)	(B,4)
C	(C,1)	(C,2)	(C,3)	(C,4)
D	(D,1)	(D,2)	(D,3)	(D,4)

Table 1.8 The Cartesian product {A, B, C, D} × {1, 2, 3, 4}.

Cartesian Products. If S is a set, and T is a set, then the Cartesian product of S and T is the set of all pairs (x, y) such that x is from S and y is from T. This is written: S × T. Notice that order matters (since we're talking about ordered pairs). So, assuming that S is not identical to T, it follows that S × T is not identical to T × S. The set of male-female pairings is not the same as the set of female-male pairings, if order matters. The pairs in Table 1.9 are distinct from those in Table 1.8.

	A	B	C	D
1	(1,A)	(2,A)	(3,A)	(4,A)
2	(1,B)	(2,B)	(3,B)	(4,B)
3	(1,C)	(2,C)	(3,C)	(4,C)
4	(1,D)	(2,D)	(3,D)	(4,D)

Table 1.9 The Cartesian product {1, 2, 3, 4} × {A, B, C, D}.

If S is any set, then we can take the Cartesian product of S with itself: S × S. Table 1.10 shows the Cartesian product of {A, B, C, D} with itself.

	A	B	C	D
A	(A,A)	(B,A)	(C,A)	(D,A)
B	(A,B)	(B,B)	(C,B)	(D,B)
C	(A,C)	(B,C)	(C,C)	(D,C)
D	(A,D)	(B,D)	(C,D)	(D,D)

Table 1.10 The Cartesian product S × S.

Just as we extended the notion of ordered pairs to ordered triples, and to ordered n-tuples generally, so also we can extend the notion of Cartesian products. If S, T, and Z are three sets, then S × T × Z is the set of all ordered 3-tuples (x, y, z) such that x is in S, y is in T, and z is in Z. Note that S × T × Z = ((S × T) × Z).

2

RELATIONS

1. Relations

Relation. A *relation* from a set S to a set T is a subset of the Cartesian product of S and T. We can put it in symbols like this: R is a relation from S to T iff (R $\subseteq$ (S $\times$ T)). A relation from a set S to a set T is a *binary relation.* Binary relations are the most common relations (at least in ordinary language). Since S $\times$ T is a set of ordered pairs, any relation from S to T is a set of ordered pairs.

For example, let Coins be the set of coins {Penny, Nickel, Dime, Quarter}. Let Values be the set of values {1, . . . 100}. The Cartesian product Coins $\times$ Values is the set of all (coin, value) pairs. One of the subsets of Coins $\times$ Values is the set {(Penny, 1), (Nickel, 5), (Dime, 10), (Quarter, 25)}. This subset of Coins $\times$ Values is a binary relation that associates each coin with its value. It's the is-the-value-of relation. Hence

is-the-value-of $\subseteq$ Coins $\times$ Values.

Domain. If R is a relation from S to T, then the *domain* of R is S. Note that the term *domain* is sometimes used to mean the set of all x in S such that there is some (x, y) in R. We won't use it in this sense.

Codomain. If R is a relation from S to T, then the *codomain* of R is T. The *range* of R is the set of all y in T such that there is some (x, y) in R. The range and codomain are not always the same. Consider the relation is-the-husband-of. The relation associates men with women. The codomain of the relation is the set of women. The range is the set of married women. Since not every woman is married, the range is not the codomain.

Of course, the domain and codomain of a relation may be the same. A relation on a set S is a subset of S $\times$ S. For example, if Human is the set of all humans, then all kinship relations among humans are subsets of the set Human $\times$ Human. As another example, the relation is-a-teacher-of is the set of all (teacher x, student y) pairs such that x is a teacher of y. Of course, we are assuming that x and y are both humans.

There are many notations for relations. If (x, y) is in a relation R, we can write xRy or R(x, y) or $x \xrightarrow{R} y$. All these notations are equivalent.

2. Some Features of Relations

Arity. We are not limited to binary relations. We can also define ternary relations. A *ternary relation* is a subset of the Cartesian product S × T × U. A *quaternary relation* is a subset of the Cartesian product S × T × U × W. And so it goes. Generally, an *n-ary relation* is a subset of $S_1 \times S_2 \times \ldots \times S_n$. An *n*-ary relation is also referred to as an *n*-place relation. The *arity* of a relation is the number of its places. So the arity of an *n*-ary relation is *n*. Note that the arity of a relation is sometimes referred to as its *degree*.

Although we are not limited to binary relations, most of the relations we use in philosophical work are binary. Relations of higher arity are scarce. So, unless we say otherwise, the term *relation* just means binary relation.

Inverse. A relation has an inverse (sometimes called a converse). The *inverse* of R is obtained by turning R around. For instance, the inverse of the relation is-older-than is the relation is-younger-than. The inverse of is-taller-than is is-shorter-than. The inverse of a relation R is the set of all (*y*, *x*) such that (*x*, *y*) is in R. We indicate the inverse of R by the symbol R^{-1}. We define the inverse of a relation R in symbols as follows:

$$R^{-1} = \{ (y, x) \mid (x, y) \in R \}.$$

Reflexivity. A relation R on a set S is *reflexive* iff for every *x* in S, (*x*, *x*) is in R. For example, the relation is-the-same-person-as is reflexive. Clark Kent is the same person as Clark Kent. All identity relations are reflexive.

Symmetry. A relation R on S is *symmetric* iff for every *x* and *y* in S, (*x*, *y*) is in R iff (*y*, *x*) is in R. For example, the relation is-married-to is symmetric. For any *x* and *y*, if *x* is married to *y*, then *y* is married to *x*; and if *y* is married to *x*, then *x* is married to *y*. A symmetric relation is its own inverse. If R is symmetric, then $R = R^{-1}$.

Anti-Symmetry. A relation R on S is *anti-symmetric* iff for every *x* and *y* in S, if (*x*, *y*) is in R and (*y*, *x*) is in R, then *x* is identical to *y*. The relation is-a-part-of is anti-symmetric. If Alpha is a part of Beta and Beta is a part of Alpha, then Alpha is identical with Beta. Note that anti-symmetry and symmetry are not opposites. There are relations that are neither symmetric nor anti-symmetric. Consider the relation is-at-least-as-old-as. Since there are many distinct people with the same age, there are cases in which *x* and *y* are distinct; *x* is at least as old as *y*; and *y* is at least as old as *x*. There are cases in which (*x*, *y*) and (*y*, *x*) are in the relation but *x* is not identical to *y*. Thus the relation is not anti-symmetric. But for any *x* and *y*, the fact that *x* is at least as old as *y* does not imply that *y* is at least as old as *x*. Hence the relation is not symmetric.

Transitivity. A relation R on S is *transitive* iff for every *x*, *y*, and *z* in S, if (*x*, *y*) is in R and (*y*, *z*) is in R, then (*x*, *z*) is in R. The relation is-taller-than is transitive. If Abe is taller than Ben, and Ben is taller than Carl, then Abe is taller than Carl.

3. Equivalence Relations and Classes

Partitions. A set can be divided like a pie. It can be divided into subsets that do not share any members in common. For example, the set {Socrates, Plato, Kant, Hegel} can be divided into {{Socrates, Plato}, {Kant, Hegel}}. A division of a set into some subsets that don't share any members in common is a *partition* of that set. Note that {{Socrates, Plato, Kant}, {Kant, Hegel}} is not a partition. The two subsets overlap – Kant is in both. More precisely, a *partition* of a set S is a division of S into a set of non-empty distinct subsets such that every member of S is a member of exactly one subset. If P is a partition of S, then the union of P is S. Thus ∪{{Socrates, Plato}, {Kant, Hegel}} = {Socrates, Plato, Kant, Hegel}.

Equivalence Relations. An *equivalence relation* is a relation that is reflexive, symmetric, and transitive. Philosophers have long been very interested in equivalence relations. Two particularly interesting equivalence relations are *identity* and *indiscernibility*.

If F denotes an attribute of a thing, such as its color, shape, or weight, then any relation of the form is-the-same-F-as is an equivalence relation. Let's consider the relation is-the-same-color-as. Obviously, a thing is the same color as itself. So is-the-same-color-as is reflexive. For any *x* and *y*, if *x* is the same color as *y*, then *y* is the same color as *x*. So is-the-same-color-as is symmetric. For any *x*, *y*, and *z*, if *x* is the same color as *y*, and *y* is the same color as *z*, then *x* is the same color as *z*. So is-the-same-color-as is transitive.

Equivalence Classes. An equivalence relation partitions a set of things into *equivalence classes*. For example, the relation is-the-same-color-as can be used to divide a set of colored things C into sets whose members are all the same color. Suppose the set of colored things C is

$$\{R_1, R_2, Y_1, Y_2, Y_3, G_1, B_1, B_2\}.$$

The objects R_1 and R_2 are entirely red. Each Y_i is entirely yellow. Each G_i is entirely green. Each B_i is entirely blue. The set of all red things in C is $\{R_1, R_2\}$. The things in $\{R_1, R_2\}$ are all color equivalent. Hence $\{R_1, R_2\}$ is one of the color equivalence classes in C. But red is not the only color. Since there are four colors of objects in C, the equivalence relation is-the-same-color-as partitions C into four equivalence classes – one for each color. The partition looks like this:

$$\{\{R_1, R_2\}, \{Y_1, Y_2, Y_3\}, \{G_1\}, \{B_1, B_2\}\}.$$

As a rule, an equivalence class is a set of things that are all equivalent in some way. They are all the same according to some equivalence relation. Given an equivalence relation R, we say the equivalence class of x under R is

$[x]_R = \{ y \mid y$ bears equivalence relation R to $x \}$.

If the relation R is clear from the context, we can just write $[x]$. For example, for each thing x in C, let the color class of x be

$[x] = \{ y \mid y \in C$ & y is the same color as $x \}$.

We have four colors and thus four color classes. For instance,

the red things = $[R_1] = [R_2] = \{R_1, R_2\}$.

We can do the same for the yellow things, the green things, and the blue things. All the things in the color class of x obviously have the same color. So

the partition of C by is-the-same-color-as = $\{ [x] \mid x \in C\}$.

Since no one thing is entirely two colors, no object can be in more than one equivalence class. The equivalence classes are all mutually disjoint. As a rule, for any two equivalence classes A and B, $A \cap B = \{\}$. Since every thing has some color, each thing in C is in one of the equivalence classes. So the union of all the equivalence classes is C. In symbols, $\cup\{ [x] \mid x \in C\} = C$. Generally speaking, the union of all the equivalence classes in any partition of any set A is just A itself.

Equivalence classes are very useful for abstraction. For instance, Frege used equivalence classes of lines to define the notion of an abstract direction (Frege, 1884: 136-139). The idea is this: in ordinary Euclidean geometry, the direction of line A is the same as the direction of line B iff A is parallel to B. The relation is-parallel-to is an equivalence relation. An equivalence class of a line under the is-parallel-to relation is

$[x] = \{ y \mid y$ is a line and y is parallel to $x \}$.

Frege's insight was that we can identify the direction of x with $[x]$. If A is parallel to B, then [A] = [B] and the direction of A is the same as the direction of B. Conversely, if the direction of A is the same as the direction of B, then [A] = [B]; hence A is in [B] and B is in [A]; so A is parallel to B. It follows that the direction of A is the direction of B if, and only if, A is parallel to B.

4. Closures of Relations

We've mentioned three important properties of relations: reflexivity, symmetry, and transitivity. We often want to transform a given relation into a relation that has one or more of these properties. To transform a relation R into a relation with a given property P, we perform the P *closure* of R. For example, to transform a relation R into one that is reflexive, we perform the reflexive closure of R. Roughly speaking, a certain way of closing a relation is a certain way of expanding or extending the relation.

Since equivalence relations are useful, we often want to transform a given relation into an equivalence relation. Equivalence relations are reflexive, symmetric, and transitive. To change a relation into an equivalence relation, we have to make it reflexive, symmetric, and transitive. We have to take its reflexive, symmetric, and transitive closures. We'll define these closures and then give a large example involving personal identity.

Reflexive Closure. We sometimes want to transform a non-reflexive relation into a reflexive relation. We might want to transform the relation is-taller-than into the relation is-taller-than-or-as-tall-as. Since a reflexive relation R on a set X contains all pairs of the form (x, x) for any x in X, we can make a relation R reflexive by adding those pairs. When we make R reflexive, we get a new relation called the *reflexive closure* of R. More precisely,

the reflexive closure of R = $R \cup \{ (x, x) \mid x \in X \}$.

For example, suppose we have the set of people {Carl, Bob, Allan}, and that Carl is taller than Bob and Bob is taller than Allan. We thus have the non-reflexive relation

is-taller-than = { (Carl, Bob), (Bob, Allan) }.

We can change this into the new reflexive relation is-taller-than-or-as-tall-as by adding pairs of the form (x, x) for any x in our set of people. (After all, each person is as tall as himself.) We thereby get the reflexive closure

is-taller-than-or-as-tall-as = { (Carl, Bob), (Bob, Allan),
(Carl, Carl), (Bob, Bob), (Allan, Allan)}.

Symmetric Closure. We sometimes want to transform a non-symmetric relation into a symmetric relation. We can change the relation is-the-husband-of into is-married-to by making it symmetric. We make a relation R symmetric by adding (x, y) to R iff (y, x) is already in R. Of course, when we make R symmetric, we get a new relation – the *symmetric closure* of R. It is defined symbolically like this:

the symmetric closure of R = $R \cup \{ (y, x) \mid (x, y) \in R \}$.

Since $\{ (y, x) \mid (x, y) \in R \}$ is the inverse of R, which we denoted by R^{-1}, it follows that

the symmetric closure of $R = R \cup R^{-1}$.

For example, suppose we have the set of people {Allan, Betty, Carl, Diane}. Within this set, Allan is the husband of Betty, and Carl is the husband of Diane. We thus have the non-symmetric relation

is-the-husband-of = { (Allan, Betty), (Carl, Diane)}.

We make this into the new symmetric relation is-married-to by taking the pairs in is-the-husband-of and adding pairs of the form (wife y, husband x) for each pair of the form (husband x, wife y) in is-the-husband-of. We thus get the symmetric closure

is-married-to = { (Allan, Betty), (Carl, Diane),
(Betty, Allan), (Diane, Carl)}.

Transitive Closure. We sometimes want to make an intransitive relation into a transitive relation. We do this by taking the *transitive closure* of the relation. The transitive closure is more complex than either the reflexive or symmetric closures. It involves many steps. We'll use the relation is-an-ancestor-of to illustrate the construction of transitive closures.

Since being an ancestor starts with being a parent, we start with parenthood. Indeed, the ancestor relation is the transitive closure of the parenthood relation. For the sake of convenience, we'll let P be the parenthood relation:

$P = \{ (x, y) \mid x \text{ is a parent of } y \}$.

Ancestors include grand parents as well as parents. The grand parent relation is a repetition or iteration of the parent relation: a parent of a parent of y is a grand parent of y. More precisely,

x is a grand parent of y iff
(there is some z)((x is a parent of z) & (z is a parent of y)).

We can put the repetition or iteration of a relation in symbols by using the notion of the *composition* of a relation with itself. It's defined for any relation R like this:

$R \circ R = \{ (x, y) \mid (\text{there is some } z)((x, z) \in R \;\&\; (z, y) \in R \}$.

The grand parent relation is obviously the composition of the parent relation with itself. In symbols, is-a-grand-parent-of = $P \circ P$. We can extend this reasoning to great grand parents like this:

x is a great grand parent of y iff
(there is some z)((x is a parent of z) & (z is a grand parent of y)).

The definition of a great grand parent is the composition of the parent relation with itself two times: is-a-great-grand-parent = $P \circ P \circ P$.

When we repeatedly compose a relation with itself, we get the *powers* of the relation:

$R^1 = R;$

$R^2 = R \circ R = R^1 \circ R;$

$R^3 = R \circ R \circ R = R^2 \circ R;$

$R^{n+1} = R^n \circ R.$

In the case of ancestor relations we have

is-a-parent-of	$= P^1$
is-a-grand-parent-of	$= P^2$
is-a-great-grand-parent-of	$= P^3$
is-a-great-great-grand-parent-of	$= P^4$.

And so it goes. We can generalize like this:

is-an-ancestor-n-generations-before $= P^n$.

We've got your ancestors defined by generation. But how do we define your ancestors? We define them by taking the union of all the generations. Your ancestors are your parents unioned with your grand parents unioned with your great grand parents and so on. Formally

is-an-ancestor-of = $P^1 \cup P^2 \cup P^3 \ldots \cup P^n \ldots$ and so on endlessly.

We said the ancestor relation is the transitive closure of the parenthood relation. And we can generalize. Given a relation R, we denote its *transitive closure* by R*. And we define the transitive closure like this:

$R^* = R^1 \cup R^2 \cup R^3 \ldots \cup R^n \ldots$ and so on endlessly.

You might object that the notion of endless unions is vague. And you'd be right. We can make it precise using numbers. Specifically, we use the *natural numbers*. These are the familiar counting numbers or whole numbers, starting with 0. And when we say *number*, without any further qualification, we mean natural number. Thus

the transitive closure of R = R* = $\cup\{ R^n \mid n \text{ is any number} \}$.

An equivalence relation is reflexive, symmetric, and transitive. So we can transform a relation R into an equivalence relation by taking its reflexive, symmetric, and transitive closures. Since we have to take three closures, there are several ways in which we can transform R into an equivalence relation. The order in which we take the symmetric and transitive closures makes a difference.

5. Recursive Definitions and Ancestrals

The transitive closure of a relation is also known as the *ancestral* of the relation. For any relation R, its ancestral is R*. We can define the ancestral of a relation by using a method known as *recursive definition*. A recursive definition involves a friendly circularity. The relation is defined in terms of itself in a logically valid way. Here's how it works with human ancestors:

x is an ancestor of *y* iff
either *x* is a parent of *y*
or there is some *z* such that *x* is a parent of *z* and *z* is an ancestor of *y*.

Observe that is-an-ancestor-of is defined in terms of itself. This sort of loop allows it to be composed with itself endlessly.

Consider the case of grand parents. If *x* is a grand parent of *y*, then there is some *z* such that

x is a parent of *z* and *z* is a parent of *y*.

The fact that *z* is a parent of *y* fits the first clause (the "either" part) of the *ancestor* definition. In other words, every parent is an ancestor. Consequently, we can replace the fact that *z* is a parent of *y* with the fact that *z* is an ancestor of *y* to obtain

x is a parent of some *z* and *z* is an ancestor of *y*.

But this fits the second clause (the "or" part) of the ancestor definition. Hence

x is an ancestor of *y*.

Consider the case of great grand parents. We have

x is a parent of z_1 and z_1 is a parent of z_2 and z_2 is a parent of *y*;

x is a parent of z_1 and z_1 is a parent of z_2 and z_2 is an ancestor of y;

x is a parent of z_1 and z_1 is an ancestor of y;

x is an ancestor of y.

The circularity in a recursive definition allows you to nest this sort of reasoning endlessly. We can do it for great great grand parents, and so on. Here's the general way to give the recursive definition of the ancestral of a relation:

x R* y iff
either x R y
or there is some z such that x R z and z R* y.

Ancestrals aren't the only kinds of recursive definitions. Recursive definition is a very useful and very general tool. We'll see many uses of recursion later (see Chapter 7). But we're not going to discuss recursion in general at this time.

6. Personal Persistence

6.1 The Diachronic Sameness Relation

One of the most interesting examples of changing an original relation into an equivalence relation can be found in the branch of philosophy concerned with *personal identity*. Since persons change considerably from youth to old age, we might wonder whether or not an older person is identical to a younger person.

A person might say that they are not the same as the person they used to be. For example, suppose Sue says, "I'm not the person I was 10 years ago". To make this statement precise, we have to make the times specific. If Sue says this on 15 May 2007, then she means Sue on 15 May 2007 is not the same person as Sue on 15 May 1997. Or a person might say that they are still the same person. Consider Anne. She might say, "Although I've changed a lot, I'm still the same person as I was when I was a little girl". She means that Anne at the present time is the same person as Anne at some past time. We could ask her to be more precise about the exact past time – what exactly does "when I was a little girl" mean? But that isn't relevant. All these statements have this form:

x at some later time t_1 is (or is not) the same person as y at some earlier time t_2.

The same form is evident in the following examples:

Sue on 15 May 2007 is not the same person as Sue on 15 May 1997;

Anne today is the same person as Anne many years ago.

When we talk about the form of an expression or statement in our text, we'll enclose it in angle brackets. This is because forms are general, and we want to distinguish between forms and instances of those forms. Thus <*x* loves *y*> is a statement form, while "Bob loves Sue" is a specific instance of that form.

An expression of the form <*x* at some later time t_1> or <*y* at some earlier time t_2> refers to a *stage* in the life of a person. It refers to a person-stage. A stage is an instantaneous slice of an object that is persisting through time. Thus you-at-2-pm is a different stage from you-at-2:01-pm. We don't need to worry about the exact ontology of person-stages here. All we need to worry about is that the relation is-the-same-person-as links person-stages that belong to (and only to) the same person. Our analysis of the relation is-the-same-person-as follows Perry (1976: 9-10). Accordingly, we say

> *x* at earlier t_1 is the same person as *y* at later t_2 iff
> the person of which *x* at t_1 is a stage
> = the person of which *y* at t_2 is a stage.

This is logically equivalent to saying

> *x* at earlier t_1 is the same person as *y* at later t_2 iff
> (there is some person P)(
> (*x* at t_1 is a stage of P) & (*y* at t_2 is a stage of P)).

As we said before, any relation of the form *x* is-the-same-F-as *y* is an equivalence relation on some set of objects. The relation is-the-same-person-as is an equivalence relation. As such, it is reflexive, symmetrical, and transitive. The personal identity relation is an equivalence relation on some set of person-stages. What set is that? Since the personal identity relation should be completely general, it has to be an equivalence relation on the set of all person-stages. We define this set like this:

> PersonStages = { *x* | (there is P)((P is a person) & (*x* is a stage of P)) }.

If our analysis of personal identity is correct, the personal identity relation will partition the set PersonStages into a set of equivalence classes. Each equivalence class will contain all and only those person-stages that belong to a single person. How can we do this?

6.2 The Memory Relation

We start with Locke's definition of a person. Locke says, "A person is a thinking intelligent being, that has reason and reflection, and can consider itself as itself, the same thinking thing, in different times and places; which it does only by that consciousness which is inseparable from thinking" (1690: II.27.9). Locke then presents his famous *memory criterion* for the persistence of a person:

> For, since consciousness always accompanies thinking, and it is that which makes every one to be what he calls self, and thereby distinguishes himself from all other thinking things, in this alone consists personal identity, i.e., the sameness of a rational being: and as far as this consciousness can be extended backwards to any past action or thought, so far reaches the identity of that person; it is the same self now it was then; and it is by the same self with this present one that now reflects on it, that that action was done. (1690: II.27.9)

According to Locke, if you can presently remember yourself as having been involved in some event, then you are the same person as the person you are remembering. For example, if you can remember yourself going to school for the first time as a child, then you are the same person as that little child. When you remember some past event in your life, you are remembering a snapshot of your life. You are remembering a stage of your life. One stage of your life is remembering another stage of your life. We now have a relation on person-stages. It is the memory relation:

x at later t_1 remembers y at earlier t_2.

Is Locke's memory criterion a good criterion? We don't care. We are neither going to praise Locke's memory criterion, nor are we going to criticize it. We're merely using it as an example to illustrate the use of set theory in philosophy. Following Locke, our first analysis of the personal identity relation looks like this:

x at later t_1 is the same person as y at earlier t_2 iff
x at later t_1 remembers y at earlier t_2.

6.3 Symmetric then Transitive Closure

As it stands, there is a fatal flaw with our initial analysis. Since the personal identity relation is an equivalence relation, it has to be symmetric. If later x is the same person as earlier y, then earlier y is the same person as later x. But our analysis so far doesn't permit symmetry. While later x remembers earlier y, earlier y can't remember later x. We can fix this problem by forming the symmetric closure of the memory relation. When we form the symmetric closure, we're defining a new relation. We can refer to it as a *psychological continuity* relation. Naturally, one might object that mere remembering isn't rich enough for real psychological continuity. But that's not our concern. We're only illustrating the formation of a symmetric closure. For short, we'll just say later x is continuous with earlier y or earlier y is continuous with later x. At this point, we can drop the detail of saying <later x> and <earlier y>. For instance, we can just say x is continuous with y or x remembers y. With all this in mind, here's the symmetric closure of the memory relation:

x is continuous with *y* iff ((*x* remembers *y*) or (*y* remembers *x*)).

If we put this in terms of sets, we have

is-continuous-with = remembers ∪ remembers^{-1}.

Our continuity relation is symmetrical. We can use it for a second try at analyzing the personal identity relation. Here goes:

x is the same person as *y* iff *x* is continuous with *y*.

As you probably expect, we're not done. There is a well-known problem with this second analysis. Continuity is based on memory. But as we all know, memory is limited. We forget the distant past. An old person might remember some middle stage of his or her life; and the middle stage might remember a young stage; but the old stage can't remember the young stage. This is known as the *Brave Officer Objection* to Locke's memory criterion. Reid (1975) gave this example: an Old General remembers having been a Brave Officer; the Brave Officer remembers having been a Little Boy; but the Old General does not remember having been the Little Boy. In this case, the memory relation looks like this:

remembers = { (Old General, Brave Officer), (Brave Officer, Little Boy) }.

We can illustrate this with the diagram shown in Figure 2.1. By the way, the diagram of a relation is also known as the *graph* of the relation.

Figure 2.1 The memory relation.

When we take the symmetric closure, we get

is-continuous-with
= { (Old General, Brave Officer), (Brave Officer, Old General),
(Brave Officer, Little Boy), (Little Boy, Brave Officer) }.

We can illustrate this with a diagram. To save space, we abbreviate is-continuous-with as continuous. Figure 2.2 shows that relation:

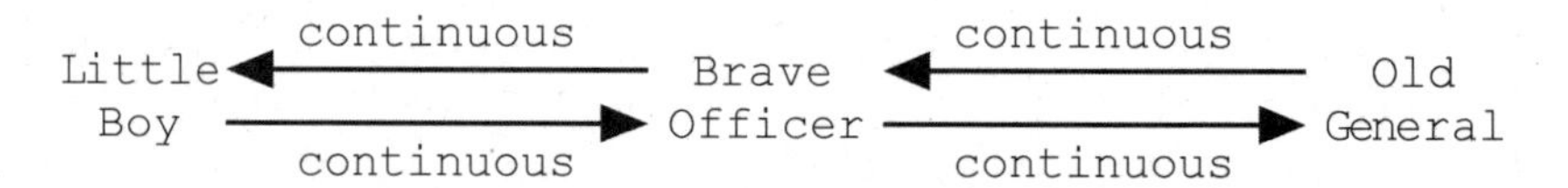

Figure 2.2 The symmetric closure of the memory relation.

We can compress the detail even further by replacing any pair of arrows with a single double-headed arrow. This compression is shown in Figure 2.3:

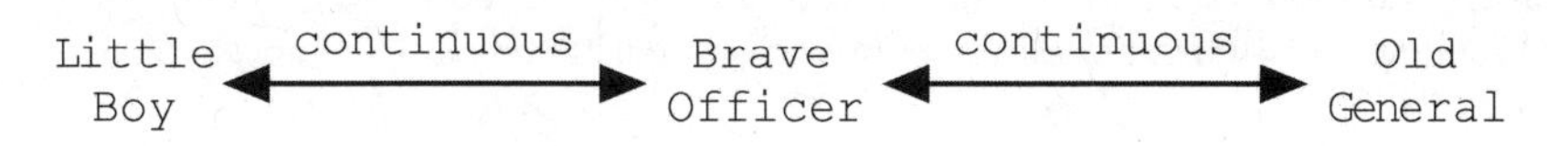

Figure 2.3 Symmetry using double-headed arrows.

The problem is that the continuity relation is not transitive. To make it transitive, we need to take the transitive closure. We need to form the ancestral of the continuity relation. Remember that R* is the ancestral of relation R. So continuous* is the ancestral of the continuity relation. Here's the ancestral:

> x is continuous* with y iff
> either x is continuous with y
> or there is some z and
> x is continuous with z and z is continuous* with y.

Consider our example. The Old General is continuous with the Brave Officer; hence, the Old General is continuous* with the Brave Officer. The Brave Officer is continuous with the Little Boy; hence the Brave Officer is continuous* with the Little Boy. Applying our definition, we see that, letting z be the Brave Officer, the Old General is continuous* with the Little Boy. Hence

> is-continuous*-with
> = { (Old General, Brave Officer), (Brave Officer, Old General),
> (Brave Officer, Little Boy), (Little Boy, Brave Officer),
> (Old General, Little Boy), (Little Boy, Old General) }.

Our diagram of this relation is shown in Figure 2.4:

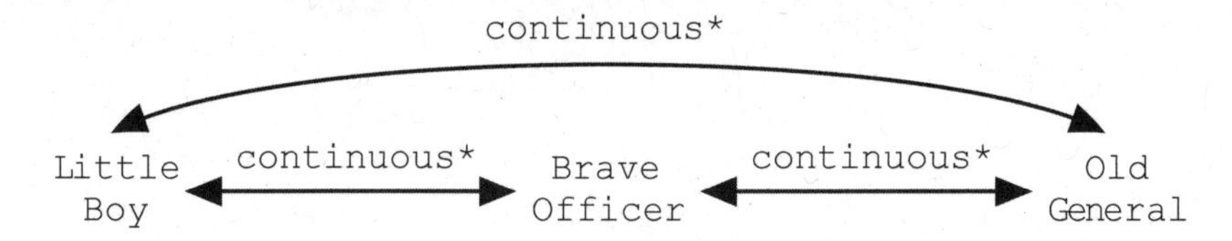

Figure 2.4 Adding the transitive closure.

We now have a relation that is based on memory, but that is both symmetrical and transitive. It is the is-continuous*-with relation. We might try to use this relation to analyze personal identity. Here goes:

> x is the same person as y iff x is continuous* with y.

Of course, this isn't quite right. We have to add reflexivity. We have to take the reflexive closure. But that's easy:

x is the same person as y iff either $x = y$ or x is continuous* with y.

In the case of our example, we need to add pairs of the form (x, x) where x is either the Old General, the Brave Officer, or the Little Boy. Here goes:

is-the-same-person-as
= { (Old General, Brave Officer), (Brave Officer, Old General),
(Brave Officer, Little Boy), (Little Boy, Brave Officer),
(Old General, Little Boy), (Little Boy, Old General),
(Old General, Old General),
(Brave Officer, Brave Officer),
(Little Boy, Little Boy) }.

Our diagram of this relation is shown in Figure 2.5:

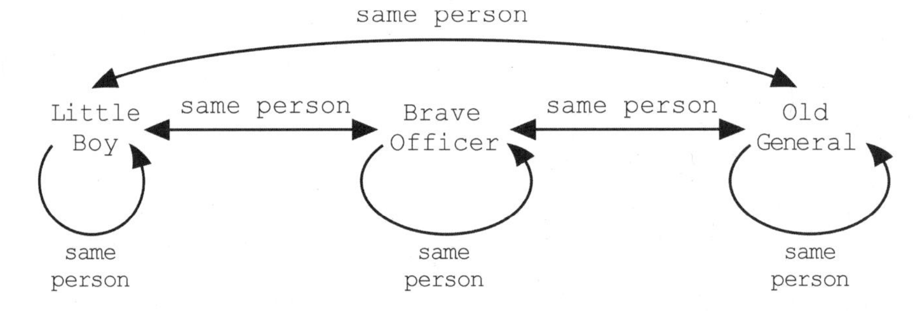

Figure 2.5 Adding the reflexive closure.

We've got a relation that's reflexive, symmetrical, and transitive. It's an equivalence relation. So we're done! Right? Not at all. Far from it. We've got an equivalence relation. But is it the correct equivalence relation? After all, you can make lots of incorrect equivalence relations on person-stages. For instance,

x is the same person as y iff x is the same height as y.

The same height relation is an equivalence relation. But nobody would say that x and y are stages of the same person iff x is exactly as tall as y! Sure, fine. Sameness of height is not the right way to analyze personal identity. But haven't we used a traditional approach? At least Locke's memory criterion is well established. What's the problem? The problem isn't that we think some other criterion is better. The problem is purely formal. The problem is revealed by cases in which persons divide. The problem is *fission*.

6.4 The Fission Problem

The fission problem was discovered by Parfit (1971). Many other writers dealt with it, notably Wiggins (1976) and Lewis (1976). We'll use it to illustrate how the order of closures makes a difference when forming equivalence relations.

We start with a set of person-stages {A, B, C, D}. Each stage remembers exactly one other stage. Stage B fissions into stages C and D. We might picture it like this: stage A walks into the front door of a duplicating machine. Once inside, stage B presses a green button labeled "Duplicate". The duplicating machine has two side doors. After a brief time, C walks out the left door and D walks out the right door. Both C and D remember having been B. They both remember pressing the green button. But they don't remember having been each other. C doesn't remember having walked out the right door and D doesn't remember having walked out the left door. Although they are similar in many ways, from the perspective of memory, C and D are total strangers. Our memory relation is

remembers = { (C, B), (D, B), (B, A) }.

The diagram of this relation is shown in Figure 2.6. An arrow from x to y displays an instance of the relation x remembers y.

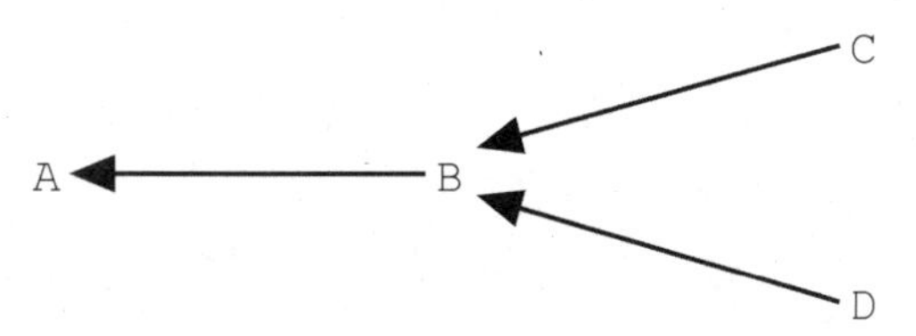

Figure 2.6 A person-process splits at stage B.

According to our earlier procedure for converting the memory relation into the personal identity relation, we took the symmetric closure, then the transitive closure, and then the reflexive closure. Let's do it in diagrams. First, we take the symmetric closure. In Figure 2.7, each arrow is the is-continuous-with relation.

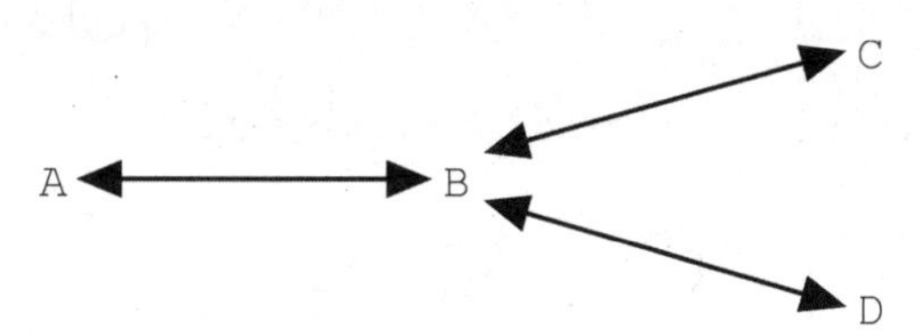

Figure 2.7 Symmetric closure.

We now take the transitive closure. Since there's an arrow from C to B, and from B to D, we have to add an arrow from C to D. And vice versa, we have to add an arrow from D to C. In fact, we have to fill in the whole diagram with arrows between any two distinct stages. We thus obtain the is-continuous*-with relation. It's displayed in Figure 2.8 like this:

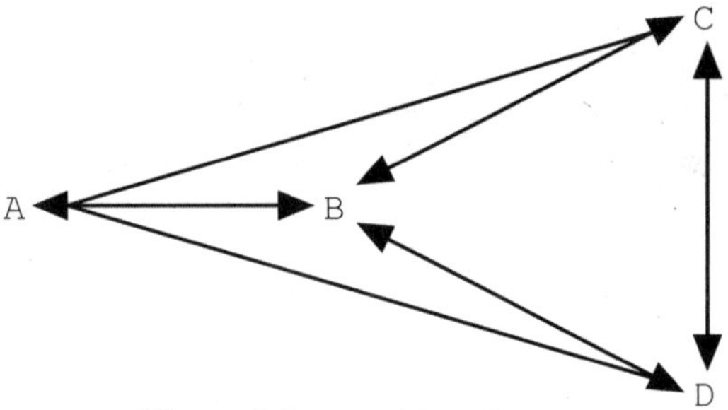

Figure 2.8 Transitive closure.

Finally, of course, we have to add the loop from each stage to itself. That makes the relation reflexive. But we don't need to show that. The result (plus the invisible loops) is the same as the diagram above. You can see that taking the symmetric and then transitive closure makes a relation in which each stage is the same person as every other stage. But this isn't right. We don't want C to be the same person as D. More generally, after fission, the stages on distinct branches should be stages of distinct persons. Taking the symmetric and then the transitive closure of the memory relation as the basis for personal identity or persistence leads to an incorrect result.

6.5 Transitive then Symmetric Closure

We want to preserve branches – that is, we want to preserve linear paths. We do this by first taking the transitive closure and then taking the symmetric closure. Finally, we take the reflexive closure. So let's work through this process.

First, we form the transitive closure of the memory relation:

> x remembers* y iff
> either x remembers y
> or there is some z such that x remembers z and z remembers* y.

Second, we take the symmetric closure of remembers* to get our continuity relation:

> x is continuous with y iff
> either x remembers* y or y remembers* x.

Finally, we take the reflexive closure:

> x is the same person as y iff
> either $x = y$
> or x is continuous with y.

This procedure works. We never add links between C and D. More generally, this procedure will never link stages on different branches after fission. It preserves the two linear chains of memories D – B – A and C – B – A. Let's look at this procedure in detail:

remembers = {(C, B), (D, B), (B, A) };

remembers* = {(C, B), (D, B), (B, A), (C, A), (D, A) };

is-continuous-with = {(C, B), (D, B), (B, A), (C, A), (D, A),
(B, C), (B, D), (A, B), (A, C), (A, D) }.

Note that continuity is well-behaved at the branch stage B. There is no cross-talk after the split. That is, the pairs (C, D) and (D, C) do *not* occur in is-continuous-with.

Finally, we take the reflexive closure:

is-the-same-person-as = { (C, B), (D, B), (B, A), (C, A), (D, A),
(B, C), (B, D), (A, B), (A, C), (A, D),
(A, A), (B, B), (C, C) }.

This procedure handles fission properly. Alas, there's still a hitch. Or at least an *apparent* hitch. Our personal identity relation isn't an equivalence relation. It can't be – the absence of any link between C and D means that it isn't fully transitive. On the contrary, it's transitive only *within* linearly ordered chains of memories. It isn't transitive *across* such chains. The two linearly ordered chains look like this:

chain-1 = { (A, B), (B, A), (A, C), (C, A), (B, C),
(C, B), (A, A), (B, B), (C, C) };

chain-2 = { (A, B), (B, A), (A, D), (D, A), (B, D),
(D, B), (A, A), (B, B), (D,D) }.

Since the relation is-the-same-person-as isn't an equivalence relation, it can't be an identity relation of any kind. Identity is necessarily reflexive, symmetric, and transitive. But is this really a problem? The is-the-same-person-as relation seems to partition person-stages into persons in exactly the right way. Within each chain, it *is* an equivalence relation. One might say that we've defined a *persistence* relation. This leads into a deep debate.

Some philosophers say that any persistence relation has to be an identity relation – after all, we're supposed to be talking about *identity* through time. For these philosophers (sometimes called *endurantists* or *3-dimensionalists*), the only solution is to say that a person ends with fission. Other philosophers say that persistence is not identity – after all, things change in time, and change negates identity. For these philosophers (sometimes called *perdurantists* or *4-dimensionalists*), our analysis of the is-the-same-person-as relation in this section is fine. It's just not an equivalence relation; hence, it isn't identity. Loux (2002: ch. 6) gives a nice overview of the debate between the endurantists and perdurantists. We won't go into it here. Our purpose has only been to show how the debate involves formal issues.

7. Closure Under an Operation

Closure under an Operation. We've discussed several operations on sets: the union of two sets; the intersection of two sets; the difference of two sets. All these operations are *binary operations* since they take *two* sets as inputs (and produce a third as output). For example, the union operator takes two sets as inputs and produces a third as output. The union of x and y is a third set z. A set S is *closed* under a binary operation ⊗ iff for all x and y in S, $x \otimes y$ is also in S.

Any set that is closed under a set-theoretic operation has to be a set of sets. Consider the set of sets S = {{A, B}, {A}, {B}}. This set is closed under the union operator. Specifically, if we take the union of {A, B} with either {A}, {B} or {A, B}, we get {A, B}, which is in S. If we take the union of {A} with {B}, we get {A, B}, which is in S. So this set is closed under union. But it is not closed under intersection. The intersection of {A} with {B} is the empty set {}. And the empty set is not a member of S.

Given any set X, the power set of X is closed under union, intersection, and difference. For example, let X be {A, B}. Then pow X is {{}, {A}, {B}, {A, B}}. You should convince yourself that pow X is closed under union, intersection, and difference. How do you do this? Make a table whose rows and columns are labeled with the members of pow X. Your table will have 4 rows and 4 columns. It will thus have 16 cells. Fill in each cell with the union of the sets in that row and column. Is the resulting set in pow X? Carry this out for intersection and difference as well.

8. Closure Under Physical Relations

An operation is a relation, so we can extend the notion of closure under an operation to closure under a relation. For example, some philosophers say that a universe is a maximal spatio-temporal-causal system of events. This means that the set of events in the universe is closed under all spatial, temporal, and causal relations.

Spatio-Temporal Closure. It is generally agreed that a universe is closed under spatial and temporal relations. For example, consider the temporal relation x is later than y. A set of events in a universe U is closed under this temporal relation. For every event x in U, for every event y, if y is later than x or x is later than y, then y is in U.

Causal Closure. Some philosophers say that universes are causally closed. This is more controversial. Jaegwon Kim defines causal closure like this: "If you pick any physical event and trace out its causal ancestry or posterity, that will never take you outside the physical domain. That is, no causal chain will ever cross the boundary between the physical and the nonphysical." (Kim, 1998: 40) Of course, these events are in some universe, presumably ours. A universe *is causally closed* iff all causes of events in the universe are in the universe, and all effects of events in the universe are in the universe. Let's take the two cases of cause and effect separately. Consider the claim that all causes of events in a universe U are in U. This means that for every event x in U, and for any event y, if y causes x, then y is in U. Consider the claim that all effects of events in a universe U are in U. This means that for every event x in U, and for any event y, if x causes y, then y is in U.

A more interesting thesis is that *physicality* is causally closed. Presumably this means that the set of physical events is closed under the relations of cause and effect. Specifically, let P be the set of physical events. To say P is causally closed is to say that all causes of physical events are physical, and all effects of physical events are physical. It means that all causes of the events in P are in P, and all effects of events in P are in P. As before, let's take the two cases of cause and effect separately. Consider the claim that all causes of a physical event are physical. This means that for any event x in P, and for any event y, if y causes x, then y is in P. Consider the claim that all effects of a physical event are physical. For any event x in P, and for any event y, if x causes y, then y is physical.

The thesis that physicality is causally closed is used in the philosophy of mind. It can be used in an argument that the mental events are physical and that the mind itself is physical. Mental events cause brain events; brain events are physical events; the set of physical events is causally closed; hence, all causes of physical events are physical events. Since mental events cause physical events (brain events), it follows that all mental events are physical events. Suppose we assume that all mental events are events involving a mind. And we assume further that if any of the events involving a thing are physical, then the thing itself is physical. On those assumptions, it follows that the mind is physical. You might accept this argument or reject some of its assumptions. We're not concerned with the objections and replies to this argument – we only want to illustrate the use of closure. An interesting parallel argument can be run using God in place of mind.

9. Order Relations

Order. A relation R on a set X is an *order relation* iff R is reflexive, anti-symmetric, and transitive. (Note that an order relation is sometimes called a *partial order*.) Since R is reflexive, for all *x* in X, (*x*, *x*) is in R. Since R is anti-symmetric, for all *x* and *y* in X, if both (*x*, *y*) and (*y*, *x*) are in R, then *x* is identical with *y*. Since R is transitive, for all *x*, *y*, and *z* in X, if (*x*, *y*) is in R and (*y*, *z*) is in R, then (*x*, *z*) is in R.

An obvious example of an order relation on a set is the relation is-greater-than-or-equal-to on the set of numbers. This relation is symbolized as $\geq$.

Quasi-Order. A relation R on X is a *quasi-order* iff R is reflexive and transitive. (Note that a quasi-order is sometimes called a *pre-order*.) Suppose we just measure age in days – any two people born on the same day of the same year (they have the same birth date) are equally old. Say *x* is at least as old as *y* iff *x* is the same age as *y* or *x* is older than *y*. The relation is-at-least-as-old-as is a quasi-order on the set of persons. It is reflexive. Clearly, any person is at least as old as himself or herself. It is transitive. If *x* is at least as old as *y*, and *y* is at least as old as *z*, then *x* is at least as old as *z*. But it is not anti-symmetric. If *x* is at least as old as *y* and *y* is at least as old as *x*, it does not follow that *x* is identical with *y*. It might be the case that *x* and *y* are distinct people with the same birth date. It's worth mentioning that not being anti-symmetric does *not* imply being symmetric. The relation is-at-least-as-old-as is neither anti-symmetric nor symmetric. For if *x* is younger than *y*, then *y* is at least as old as *x* but *x* is not at least as old as *y*.

The difference between order relations and quasi-order relations can be subtle. Consider the relation is-at-least-as-tall-as. Suppose this is a relation on the set of persons, and that there are some distinct persons who are equally tall. The relation is-at-least-as-tall-as is reflexive. Every person is at least as tall as himself or herself. And it is transitive. However, since there are some distinct persons who are equally tall, it is not anti-symmetric. So it is not an order relation. It is merely a quasi-order.

But watch what happens if we restrict is-at-least-as-tall-as to a subset of people who all have distinct heights. In this subset, there are no two people who are equally tall. In this case, is-at-least-as-tall-as remains both reflexive and transitive. Now, since there are no two distinct people *x* and *y* who are equally tall, it is always true that if *x* is at least as tall as *y*, then *y* is not at least as tall as *x*. For if *x* is at least as tall as *y*, and there are no equally tall people in the set, then *x* is taller than *y*. Consequently, it is always false that ((*x* is at least as tall as *y*) & (*y* is at least as tall as *x*)). Recall that if the antecedent of a conditional is false, then the conditional is true by default. So the conditional statement (if ((*x* is at least as tall as *y*) & (*y* is at least as tall as *x*)) then $x = y$) is true by default. So is-at-least-as-tall-as is anti-symmetric on any set of people who all have distinct heights. Therefore, is-at-least-as-tall-as is an order relation on any set of people who all have distinct heights.

10. Degrees of Perfection

A long tradition in Western thought treats reality as a *great chain of being* (Lovejoy, 1936). The chain is a series of levels of perfection. As you go higher in the levels, the things on those levels are increasingly perfect. Some philosophers used this reasoning to argue for a maximally perfect being at the top of the chain. The argument from degrees of perfection is also known as the *Henological Argument*. An early version of the Henological Argument was presented by Augustine in the 4th century (1993: 40-64). Anselm presented his version of the Henological Argument in Chapter 4 of the *Monologion*. Aquinas presented it as the Fourth Way in his Five Ways (Aquinas, *Summa Theologica*, Part 1, Q. 2, Art. 3). Since Anselm's version is the shortest and sweetest, here it is:

> if one considers the natures of things, one cannot help realizing that they are not all of equal value, but differ by degrees. For the nature of a horse is better than that of a tree, and that of a human more excellent than that of a horse . . . It is undeniable that some natures can be better than others. None the less reason argues that there is some nature that so overtops the others that it is inferior to none. For if there is an infinite distinction of degrees, so that there is no degree which does not have a superior degree above it, then reason is led to conclude that the number of natures is endless. But this is senseless . . . there is some nature which is superior to others in such a way that it is inferior to none. . . . Now there is either only one of this kind of nature, or there is more than one and they are equal . . . It is therefore quite impossible that there exist several natures than which nothing is more excellent. . . . there is one and only one nature which is superior to others and inferior to none. But such a thing is the greatest and best of all existing things. . . . there is some nature (or substance or essence) which is good, great, and is what it is, through itself. And whatsoever truly is good, great, and is a thing, exists through it. And it is the topmost good, the topmost great thing, the topmost being and reality, i.e., of all the things that exist, it is the supreme. (Anselm, 1076: 14-16)

We are not at all interested in whether Anselm's Henological Argument is sound. We are interested in formalizing it in order to illustrate the use of set theory in philosophy. For Anselm's argument to work, we have to be able to compare the perfections of things. We assume that things are the individuals in our universe – that is, they are the non-sets in our universe. For Anselm, humans are more perfect than horses; horses are more perfect than trees. But while Anselm gives these examples, he isn't very clear about exactly what is more perfect than what. We need some general principles.

The idea of a hierarchy of degrees of perfection is not original with Anselm. It's an old idea. Long before Anselm, Augustine outlined a hierarchy of degrees of perfection (his ranking is in *The City of God*, bk. XI, ch. 16). And Augustine's hierarchy is a bit clearer. He distinguishes five degrees of perfection: (1) merely existing things (e.g., rocks); (2) living existing things (e.g., plants); (3) sentient living existing things (e.g., animals); (4) intelligent sentient living things (e.g., humans); (5) immortal sentient intelligent living things (angels). Now suppose the set of things in the universe is:

{ theRock, theStone, thePebble,
theTree, theBush, theFlower, theGrass,
thePuppy, theCat, theHorse, theLion,
Socrates, Plato, Aristotle,
thisAngel, thatAngel }.

We start with the equivalence relation is-as-perfect-as. Following Augustine, though not exactly on the same path, we'll say that any two rocks are equally perfect; any two plants are equally perfect; and so on for animals, humans, and angels. Recall that we use $[x]_R$ to denote an equivalence class under R. Here we know that R is just is-as-perfect-as; so we can just write $[x]$. We define the perfection class of a thing as expected:

$[x] = \{ y \mid y$ is as perfect as $x \}$.

For example, our sample universe contains just three merely existing things: theRock, theStone, and thePebble. These are all equally perfect. Hence

[theRock] = { $y \mid y$ is as perfect as theRock }.

Spelling this out, we get

[thcRock] = { thcRock, thcStonc, thcPcbblc }.

Since theRock, theStone, and thePebble are all equally perfect, they are all in the same degree of perfection. For each of these things, the set of all equally perfect things contains exactly theRock, theStone, and thePebble. Thus

[theRock] = [theStone] = [thePebble].

We now partition the set of things into perfection classes. These are equivalence classes. These are the *degrees of perfection*. Based on Augustine's ranking of things in terms of their perfections, our sample universe divides into these degrees of perfection:

D_1 = { theRock, theStone, thePebble };
D_2 = { theTree, theBush, theFlower, theGrass };
D_3 = { thePuppy, theCat, theHorse, theLion };
D_4 = { Socrates, Plato, Aristotle };
D_5 = { thisAngel, thatAngel }.

The set of such degrees is

degrees-of-perfection = {[x] | x is a thing }.

Which in this case is

degrees-of-perfection = { D_1, D_2, D_3, D_4, D_5 }.

So far we've been dealing with the comparative perfection relations between things. But we can extend those relations to degrees of perfection, that is, to sets of things. Given any two degrees of perfection X and Y, we say

X is higher than Y iff every x in X is more perfect than any y in Y.

In our example, D_5 is higher than D_4, D_4 is higher than D_3, and so on. The relation is-a-higher-degree-than is a comparative relation. It is transitive. We can make it reflexive by taking the reflexive closure:

X is at least as high as Y iff X is higher than Y or X is identical with Y.

The relation is-at-least-as-high-as is reflexive and transitive. So it is a quasi-order on the set of degrees of perfection. However, since no two degrees are equally perfect, this quasi-order is anti-symmetric by default. Hence it is an order relation.

We thus obtain an ordered set of degrees of perfection. Anselm thinks that the degrees of perfection have a simple finite ordering. There is a lowest degree of perfection. There is a finite series of higher degrees. Above them all, there is a top degree of perfection. This degree contains one thing: the maximally perfect being, God. Following Augustine's ordering, it would appear that God exists in the 6th degree:

D_6 = { God }.

At this point, you should have a lot of questions. Can't the series of degrees rise higher endlessly like the numbers? Can God be a member of a set? Does this mean sets are independent of God? Interesting questions. What are your answers?

11. Parts of Sets

Mereology is the study of parts and wholes. It is an important topic in contemporary formal philosophy. Although mereology was for some time thought of as an alternative to set theory, Lewis (1991) argued that mereology and set theory are closely interrelated. To put it crudely, you can do mereology with sets. And, conversely, there are ways to do something similar to set theory using mereology. Lewis's work on the relations between set theory and mereology is deep and fascinating. We can't go into it here. Here, we just discuss the parts of sets. But first, we talk about parts and wholes.

It often happens that one thing is a part of another. For instance, the wheel is a part of the bicycle. And we can even think of a thing as a part of itself. Hence the bicycle is a part of the bicycle. To be clear, we distinguish between proper and improper parts, just as we distinguished between proper and improper subsets. For any x and any y, x is an improper part of y iff x is identical with y. In other words, every thing is an improper part of itself. For any x and any y, x is a proper part of y iff x is a part of y and x is not identical with y. The wheel is a proper part of the bicycle; the bicycle is an improper part of itself. We'll symbolize the is-a-part-of relation by $<<$. It's defined by three rules:

1. The parthood relation is reflexive. Everything is a part of itself. For all x, $x << x$.

2. The parthood relation is anti-symmetric. If two things are parts of each other, then they are identical. For all x and for all y, if $x << y$ and $y << x$, then $x = y$.

3. The parthood relation is transitive. For any x, y, and z, if $x << y$ and $y << z$, then $x << z$.

As we look through our set-theoretic relations, we can see that the subset relation has the same logical form as the parthood relation – the subset relation is also reflexive, anti-symmetric, and transitive. Indeed, we can think of the subsets of a set as the parts of the set. Here's how we do it. Start with a set S. Now take the power set of S. The power set is the set of all subsets of S. You might think that the power set is the set of all parts of S. But the power set includes the empty set, and the empty set doesn't have any content. So we exclude it from the set of parts of S. For any set S, we define

the parts of S = pow S – $\{\{\}\}$.

And for any x and y in the parts of S, we say x is a part of y iff x is a subset of y. In symbols, we say

$x << y$ iff $x \subseteq y$.

For example, if S = {A, B, C}, then

the parts of S = {{A}, {B}, {C}, {A, B}, {A, C}, {B, C}, {A, B, C}}.

Using the subset relation, we see that {A} is a part of {A, B}, and that {A, B} is a part of {A, B, C}. Further, {A, B, C} is a part of {A, B, C}, but it is an improper part.

You can see immediately that some parts of S have no further parts. They are partless. For the mereologist, parts that have no further parts are *atoms*. As a rule, the unit sets are the atomic parts of a set. That is, x is an atomic part of S iff there is some y in S and x is the unit set of y. Formally,

x is an atomic part of S iff there is some $y \in$ S and $x = \{y\}$.

Thus {A} is an atomic part of {A, B, C}, as are {B} and {C}. We can now define various mereological relations in set-theoretic terms. For example, mereologists say

x overlaps y = there is some z such that z is a part of x and z is a part of y.

And we can define this set-theoretically as

x overlaps y = there is some z such that $z \in (x \cap y)$.

In other words, x overlaps y iff they have a non-empty intersection. For example, {A, B} overlaps {B, C} because they both have {B} as a part. We could go on to translate other mereological relations into set-theoretic terms, but we leave that up to you. You might, for instance, work on translating Chapter 3 in Casati & Varzi (1999) into the language of sets.

12. Functions

Function. A function f from set X to set Y is a relation in which each member of X has a unique partner in Y. Functions are special because they are *unambiguous*. Each member x in X is paired off with *exactly one* member y of Y. We symbolize a function from X to Y as f: X $\rightarrow$ Y. If a function f from X to Y pairs off some x in X with some y in Y, we say f of x is y, and we write this as $f(x) = y$. Thus for any x in X, the symbolism $f(x)$ refers to the unique partner of x in Y.

For example, a *seating assignment* is a function f from some set of Students to some set of Desks. It is thus a function f: Students $\rightarrow$ Desks. A seating assignment pairs off every student in Students with exactly one desk in Desks. If Susie is a member of Students, then f(Susie) is the desk at which Susie sits. The function f pairs Susie with Susie's desk.

For example, a *grade function* is a function G from the set of Students in some class to a set of Grades. It is a function G: Students → Grades. Suppose G is the grading function for a Metaphysics course. Function G pairs off every student with his or her unique grade. After all, a student can't get more than one grade in the same course. For example, if Sam gets an A in Metaphysics, then G(Sam) = A.

If f is a function from X to Y, we refer to X as the *domain* of f and Y as the *codomain* of f. We say function f *maps* its domain to its codomain or that it *is a map from* its domain *to* its codomain. The seating assignment function maps Students to Desks. The grading function maps Students to Grades.

A function f from X to Y is said to *assign* a member of Y to each member of X. If $f(x) = y$, then f assigns y to x. The seating assignment function f assigns a desk to Susie. The grade function G assigns a grade to each student in Metaphysics.

A function does *not* pair a member of X with more than one member of Y. This is why functions are singled out for special focus. They are *unambiguous* relations. A seating assignment function does not pair a student with more than one desk. The seating assignment *uniquely determines* your seat. There is no ambiguity or confusion about your seat. A grading function does not pair a student with more than one grade. The grading function uniquely determines your grade; there is no ambiguity.

A function does not *fail* to pair a member of X with some member of Y. A seating assignment function does not fail to assign a desk to a student. Every student is partnered with a desk. Every student gets to sit down. A grading function does not leave any student without some grade. Every student is paired off with a grade.

Many-to-One. Although a function from X to Y *cannot* associate one member of X with many members of Y, it *can* associate one member of Y with many members of X. One student cannot get many grades, but one grade can be given to many students. Both Sam and Susie may get As in Metaphysics. If that happens, then G(Sam) = G(Susie). If that happens, the grading function is many-to-one. Many students get one grade.

One-to-One. A function is *one-to-one* if, and only if, it never assigns one member of Y to more than one member of X (though it may leave some members of Y unassigned). For example, the seating assignment function is one-to-one. One student gets one desk, and one desk gets one student (though if there are more desks than students, some desks may get no students). Since the seating assignment is one-to-one, if the desk assigned to Superman = the desk assigned to Clark Kent, then you can infer that Superman = Clark Kent. But the grading function need not be one-to-one. One student gets one grade, but one grade can be given to many students. If the grade assigned to Susie = the grade assigned to Susan, you cannot infer that Susie = Susan. They could be distinct. We can express this symbolically like this: a function f is one-to-one if, and only if, $f(x) = f(y)$ implies $x = y$. A function that is one-to-one is also said to be 1-1. (Note that some authors use *1-1* and *one-to-one* differently, but, for us, they mean the same.)

Although a function from X to Y cannot leave any member of X without a partner in Y, it can leave some members of Y without partners in X. It may happen that a small class meets in a large classroom. If there are more desks than students, then a seating assignment function will pair each student with a desk, but it will not pair every desk with a student. Some desks will be empty. It may happen that all the students in a class pass the class. Yay! If every student in Metaphysics passes, then the grade F is not used. Although every student gets partnered with a grade, not every grade gets partnered with a student. Since every student passes, the failing grade F is not assigned to any student.

Onto. A function from X to Y is either a function *onto* Y or *into* Y. A function from X to Y is onto if and only if it associates every member of Y with some member of X. No members of Y are left without partners in X. For example, if there are exactly as many desks in a classroom as there are students in that class, then every student gets one desk and every desk gets one student. Hence the relation f that partners students with desks is a function from Students *onto* Desks. The fact that every student gets one desk makes f a function; the fact that every desk gets one student makes f onto. More formally, a function f is onto iff for every $y \in$ Y, there is some $x \in$ X such that $f(x) = y$. A function that is not onto is into. For example, if there are more desks than students in some class, then the seating assignment is a function from the set of students *into* the set of desks.

Over the years, various alternative terms have evolved for different kinds of functions. A one-to-one function is sometimes said to be an *injection*. An onto function is sometimes said to be a *surjection*. And a function that is both one-to-one and onto is sometimes called a *bijection*. It is also sometimes known by the term *1-1 correspondence*. The Figures 2.9A through 2.9D show four kinds of functions: (9A) one-to-one and onto; (9B) one-to-one and not onto; (9C) many-to-one and onto; and (9D) many-to-one and not onto.

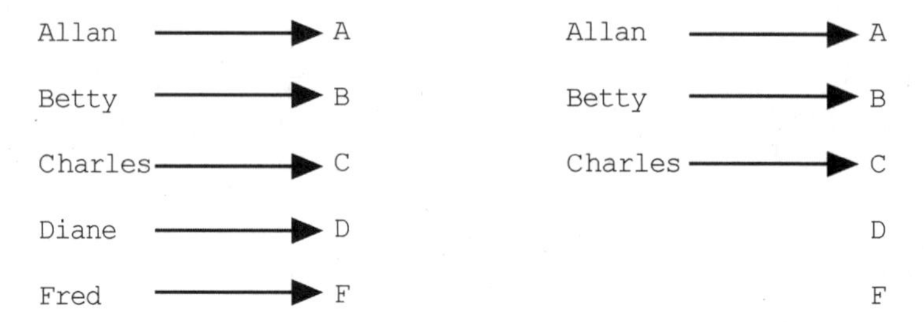

Figure 2.9A One-to-one and onto.

Figure 2.9B One-to-one, not onto.

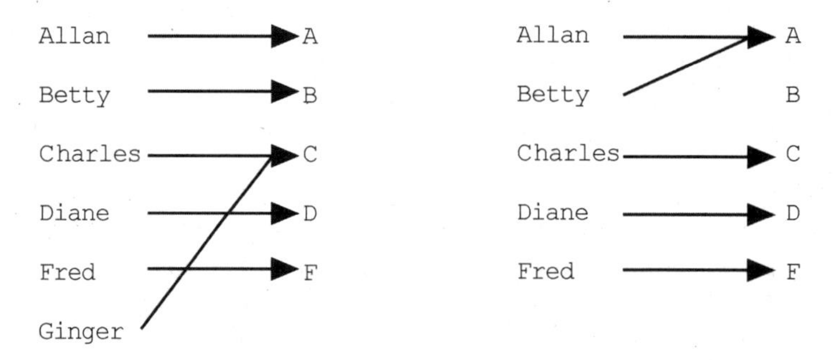

Figure 2.9C Many-to-one and onto.

Figure 2.9D Many-to-one, not onto.

Inverse. A function has an *inverse*. The inverse of f is denoted f^{-1}. The inverse of a function is defined like this: $f^{-1} = \{ (y, x) \mid (x, y) \in f \}$. If a function is one-to-one, then its inverse is also a function. If a function is many-to-one, then its inverse is not a function.

Consider the function from students to grades in Figure 2.9B. Writing it out, this function is {(Allan, A), (Betty, B), (Charles, C)}. Since this function is 1-1, the inverse of this function is also a function. The inverse of this function is the function {(A, Allan), (B, Betty), (C, Charlie)}. Now consider the function in Figure 2.9D. The inverse of this function is the relation {(A, Allan), (A, Betty), (C, Charles), (D, Diane), (F, Fred)}. Since this relation associates A with both Allan and Betty, it isn't a function.

13. Some Examples of Functions

We said that a function is an *unambiguous* relation. Consider reference. Many philosophers believe that there is a reference relation that associates names with things. A reference relation is ambiguous if it associates one name with many things. For example, if "Jim" refers to Big Jim and Little Jim, then the reference of "Jim" is ambiguous. But a reference relation is unambiguous if it associates one name with exactly one thing. If a reference relation is unambiguous, then it is a reference function. Of course, a reference function can associate many names with one thing. For example, a reference function can associate both "Superman" and "Clark Kent" with the Man of Steel.

We can think of a language as a set of sentences. Accordingly, the English language is a set of sentences. Let this set be L. Sentences are often associated with truth-values. The sentence "Snow is white" is true while the sentence "Dogs are reptiles" is false. The set of truth-values is {T, F}. Suppose Bob is a philosopher who argues that every sentence in the English language is exclusively either true or false. No sentence lacks a truth-value and no sentence is both true and false. What is Bob arguing for? He is arguing that there is a *truth-value assignment* f from L to {T, F}. Since each sentence in L has a unique truth-value, f is a function. Since there are many more sentences than truth-values, f is many-to-one. Since some sentences are true while others are false, f is onto.

Functions are useful in science (and thus in philosophy of science) for assigning attributes to things. For example, every subatomic particle has a unique charge. An electron has a charge of –1; a neutron has a charge of 0; a proton has a charge of +1. Let Particles be the set of all electrons, protons, or neutrons in our universe. Let Charges be {-1, 0, +1}. The charge function q is a map from Particles to Charges. Thus

$$\text{q: Particles} \rightarrow \text{Charges.}$$

The charge function is many-to-one and onto.

Functions can map pairs (or longer n-tuples) onto objects. For example, consider the set of all pairs of humans. This set is Humans × Humans. For every pair of humans, we can define the length of time in days that they've been married. If two humans are not married, we say that the length of time they've been married is 0. The function M associates every pair (human, human) with a number. Hence M is a set of ordered pairs of the form ((human, human), days). Letting N be the set of natural numbers, the form of M is:

$$\text{M: Humans} \times \text{Humans} \rightarrow \text{N.}$$

Suppose we let K = ((Humans × Humans) × N). Each item in M is a member of K. That is, M is a subset of K. And we can specify M more precisely as

$$\text{M} = \{((x, y), z) \in \text{K} \mid x \text{ and } y \text{ have been married for } z \text{ days}\}.$$

It's often convenient to display a function in a table. For a function like M, the table is a matrix. It has rows and columns. For each human x in Humans, there is a row for x. And, for each human y in Humans, there is a column for y. Consequently, for each couple (x, y) in Humans × Humans, there is a cell in the matrix. The cell is located at row x and column y. Suppose our sample set of humans is

Humans = {Bob, Sue, Linda, Charles}.

Table 2.1 shows the marital relations of these humans. Take the case of Bob and Sue. They got married yesterday. Hence M(Bob, Sue) = 1. In other words ((Bob, Sue), 1) is a member of M. Likewise ((Sue, Bob), 1) is in M. But Linda and Charles are not married. Hence for any y in Humans, M(Linda, y) is 0 and M(Charles, y) is 0.

	Bob	Sue	Linda	Charles
Bob	0	1	0	0
Sue	1	0	0	0
Linda	0	0	0	0
Charles	0	0	0	0

Table 2.1 The length of some marriages.

As another example, consider distances between cities. The set of cities is

Cities = {New York, Chicago, San Francisco}.

The distance function D associates each pair of cities with a number (say, the number of whole miles for driving between the cities). Recall that N is the natural numbers. So

D: Cities $\times$ Cities $\rightarrow$ N.

Each member of D is a pair ((city, city), miles). Thus

D = { ((city x, city y), miles m) | the distance between x and y is m }.

Table 2.2 illustrates the distances between the cities in Cities.

	New York	Chicago	San Francisco
New York	0	791	2908
Chicago	791	0	2133
San Francisco	2908	2133	0

Table 2.2 Distances between some cities.

Characteristic Function. A characteristic function is a function f from some set S onto {0, 1}. A characteristic function over a set S is a way of specifying a subset of S. For any x in S, x is in the subset specified by f if $f(x) = 1$ and x is not in that subset if $f(x) = 0$.

For example, consider the set {A, E, I, O, U, Y}. Now consider the function C = { (A, 0), (E, 1), (I, 1), (O, 1), (U, 0), (Y, 0)}. The function C is a characteristic function. It specifies the subset {E, I, O}. Consider the function D = { (A, 1), (E, 0), (I, 0), (O, 0), (U, 1), (Y, 1)}. D is a characteristic function that corresponds to the subset {A, U, Y}.

We can use characteristic functions to introduce the idea of a set of functions. The set of characteristic functions over a set S is the set of all f such that f: S → {0, 1}. In symbols,

the set of characteristic functions over S = { f | f: S → {0, 1}}.

Since each characteristic function over S specifies a subset, there is a 1-1 correspondence between the characteristic functions over S and the power set of S. When we were talking about defining sets as selections (given by truth-tables), we were thinking of sets in terms of characteristic functions.

14. Isomorphisms

An *isomorphism* is a structure-preserving bijection – it is a 1-1 correspondence between two systems that exactly translates the structure of the one into the structure of the other. Thus two systems that are isomorphic have the same structure or the same form. Isomorphism is used in many ways in philosophy. For example, some philosophers say that a thought or statement is true if, and only if, the structure of the thought or statement corresponds to the structure of some part of reality. Usually, the correspondence is said to be an isomorphism. Isomorphism is therefore important in theories of meaning and mental representation.

A good initial example of an isomorphism is the correspondence between the four compass directions and the four seasons. There are two sets:

theSeasons = {winter, spring, summer, fall};
theDirections = {north, south, east, west}.

Each set is structured by two relations: opposition and orthogonality. North and south are opposites; north and west are orthogonal. Winter and summer are opposites; winter and spring are orthogonal. Table 2.3 illustrates the relations that structure each set.

opposite(north, south)	opposite(winter, summer)
opposite(east, west)	opposite(spring, fall)
orthogonal(north, east)	orthogonal(winter, spring)
orthogonal(north, west)	orthogonal(winter, fall)
orthogonal(south, east)	orthogonal(summer, spring)
orthogonal(south, west)	orthogonal(summer, fall)

Table 2.3 The structure of theDirections and theSeasons.

Given these structures, there are many possible isomorphisms. One of these isomorphisms is shown in Table 2.4. Call this isomorphism f. It is a function from theSeasons to theDirections. Thus f: theSeasons → theDirections.

winter	→ north
spring	→ east
summer	→ south
fall	→ west

Table 2.4 An isomorphism f from theSeasons to theDirections.

You should see that f preserves the structure of the two sets. Since f preserves structure, it can be used to *translate* facts about seasons into facts about directions. Suppose some relation R holds between two seasons X and Y. Since f preserves structure, the same relation holds between the directions f(X) and f(Y). For example, since winter and summer are opposites, we know that f(winter) and f(summer) are also opposites. Now, f(winter) is north and f(summer) is south, and indeed north and south are opposites. More generally, for any seasons x and y, and for any structuring relation R,

R(x, y) if and only if R(f(x), f(y)).

Another good example is a map and its territory. Suppose M is a map of California. Towns are indicated by labels linked by lines. If M is accurate, then the distance relations between labels on M mirror the distance relations between cities in California. This mirroring is isomorphism. More precisely, the map-territory relation between labels and towns is an isomorphism iff the distance between labels mirrors the distance between towns. Of course, there is some proportionality (some scale of distance). Thus 1 inch on the map may be 1 mile in California – it's a big map. The map-territory relation is an isomorphism iff for any labels x and y on the map, the distance in inches between x and y is proportional to the distance in miles between the town represented by x and the town represented by y.

We can define the isomorphism between map and territory more formally. First, let the map-territory relation be f. Note that f is in fact a function. It associates each label with exactly one town. For example, f("Sacramento") = Sacramento and f("Los Angeles") = Los Angeles. We say the map-territory function f is an isomorphism iff for any labels x and y on the map, the distance between x and y is proportional to the distance between the town f(x) and the town f(y). For any labels x and y, let the distance in inches between them on the map be d(x, y). For any towns X and Y, let the distance in miles between them on the ground be D(X, Y). Suppose label x represents town X and label y represents town Y. Thus d(x, y) in inches = D(X, Y) in miles. But the representation relation is f. So f is an isomorphism iff for any x and y on the map, d(x, y) in inches = D(f(x), f(y)) in miles. Figure 2.10 illustrates the isomorphism for Sacramento and Los Angeles.

Figure 2.10 The map-territory relation for some labels and cities.

To help see the nature of an isomorphism, consider spatial relations like is-east-of and is-to-the-right-of. Suppose our map is laid out with north at its top and east at its right. Given such a layout, for any labels x and y, x is to the right of y iff the city named by x is to the east of the city named by y. Now, the city named by x is $f(x)$ and the city named by y is $f(y)$. Thus x is to the right of y iff $f(x)$ is to the east of $f(y)$. Letting R be is-to-the-right-of and E be is-to-the-east-of, we can say $\mathrm{R}(x, y)$ iff $\mathrm{E}(f(x), f(y))$.

Isomorphism. Consider two systems X and Y. The system X is the pair (A, R) where A is a set of objects and R is a relation on A. The system Y is (B, S) where B is a set of objects and S is a relation on B. A function f from A to B is an isomorphism from X to Y iff it preserves the relational structure. More precisely, the function f is an *isomorphism* iff for any x and y in A, $\mathrm{R}(x, y)$ is equivalent to $\mathrm{S}(f(x), f(y))$.

Isomorphisms have been used many times in philosophy. Here we give two examples: (1) a universe with infinite two-way eternal recurrence; (2) Burks' dual universe (1948-49: 683). These isomorphisms give us our first exposure to the notion of a *counterpart*.

An eternally recurrent universe involves a single kind of cosmic period that is endlessly repeated both into its past and into its future. Every eternally recurrent universe contains infinitely many isomorphic parts: each instance of the repeated cosmic period is isomorphic to every other instance. Each instance of the cosmic period exactly mirrors every other instance. The ancient Greeks talked about eternal recurrence. For example, the ancient Greek philosopher Eudemus stood before his students and said: "If one were to believe the Pythagoreans, with the result that the same individual things will recur, then I shall be talking to you again sitting as you are now, with this pointer in my hand, and everything else will be just as it is now". (Kirk & Raven, 1957: Frag. 272)

Suppose the universe is eternally recurrent, and that we're currently in the n-th repetition of the cosmic period (the n-th cycle). The current Eudemus is the n-th Eudemus; the next Eudemus is the $(n+1)$-th Eudemus. They are exact duplicates. Suppose further that f maps every individual x_n in cosmic period n

onto the individual x_{n+1} in the next cosmic period $n+1$. For example, f maps Eudemus$_n$ onto Eudemus$_{n+1}$. The function f is an isomorphism. For any relation R that holds between x_n and y_n, the same relation R holds between $f(x_n)$ and $f(y_n)$. Thus if person x_n holds pointer y_n in this cosmic cycle, then person $f(x_n)$ holds pointer $f(y_n)$ on the next cosmic cycle. So the isomorphism f preserves all the relations from cycle to cycle. It preserves the structure of the cycles. For every individual x in any eternally recurrent universe, *$f(x)$ is the recurrence counterpart of x*. Table 2.5 shows some facts about two cosmic cycles and the isomorphism that maps the one cycle onto the next.

Current Cycle	**Next Cycle**	**Isomorphism**
E is a teacher	E* is a teacher	E → E*
P is a pointer	P* is a pointer	P → P*
S is a student	S* is a student	S → S*
E holds P	E* holds P*	T → T*
E teaches S	E* teaches S*	
E teaches T	E* teaches T*	
S sits beside T	S* sits beside T*	

Table 2.5 Parts of two isomorphic cosmic cycles.

Burks (1948-49: 683) describes universes that have an internal spatial symmetry. Black (1952) later popularized the notion of a *dual universe*. A dual universe is split into two halves. The two halves are eternally synchronized (like two copies of the same movie playing simultaneously). Black says: “Why not imagine a plane running clear through space, with everything that happens on one side of it always exactly duplicated at an equal distance on the other side. . . A kind of cosmic mirror producing real images” (p. 69). Each person on the one side of the cosmic mirror has a counterpart on the other side: the Battle of Waterloo occurs on each side, “with Napoleon surrendering later in two different places simultaneously” (p. 70). Every event takes place at the same time on both the left and right sides of the cosmic mirror. For example, if the Napoleon on the one side marries the Josephine on that side, then the Napoleon on the other side marries the Josephine on the other side. Duplicate weddings take place, with duplicate wedding guests, wedding cakes, and so on.

The two sides of Black's universe have the same structure for all time. Each individual on one side has a partner on the other. The isomorphism f associates each thing on the one side with its *dual counterpart* on the other. For any things x and y, and for any relation R, R(x, y) holds on the one side iff R($f(x)$, $f(y)$) holds on the other. Table 2.6 shows parts of the two duplicate sides of the universe and their isomorphism.

This Side	That Side	Isomorphism
N is Napoleon	N* is Napoleon*	$N \rightarrow N^*$
J is Josephine	J* is Josephine*	$J \rightarrow J^*$
W is Waterloo	W* is Waterloo*	$W \rightarrow W^*$
E is Elba	E* is Elba*	$E \rightarrow E^*$
N marries J	N* marries J*	
N surrenders at W	N* surrenders at W*	
N is exiled to E	N* is exiled to E*	

Table 2.6 Parts of two isomorphic cosmic halves.

15. Functions and Sums

One advantage of functional notation is that we can use it to express the sum of some quality of the members of A. Suppose the objects in A are not numbers, but ordinary physical things. These things have weights. The function WeightOf maps Things to Numbers. The sum of the weights of the things in A is

$$\text{the sum, for all } x \text{ in A, of the weight of } x = \sum_{x \in A} WeightOf(x).$$

Sums over qualities are sometimes used in ethics. For instance, a utilitarian might be interested in the total happiness of a world. One way to express this happiness is to treat it as the sum, for all x in the world, of the happiness of x. Thus

$$\text{the happiness of world } w = \sum_{x \in w} HappinessOf(x).$$

We can nest sums within sums. For example, suppose each wedding guest x gives a set of gifts GIFTS(x) to the bride and groom. Each gift y has a value V(y). The total value of the gifts given by x is the guest value GV(x). It is symbolized like this:

$$\text{GV}(x) = \text{the sum, for all } y \text{ in GIFTS}(x)\text{, of V}(y) = \sum_{y \in \text{GIFTS}(x)} \text{V}(y).$$

Hopefully, our dear bride and groom have many guests at their wedding. Suppose the set of wedding guests is GUESTS. The total value of gifts given to the couple is TV. We define TV like this:

$$\text{TV} = \text{the sum, for all } x \text{ in GUESTS, of GV}(x) = \sum_{x \in \text{GUESTS}} \text{GV}(x).$$

We can, of course, substitute the definition of GV(x) for GV(x). We thus get a *nested sum*, or a sum within a sum. Here it is:

$$\text{TV} = \sum_{x \in \text{GUESTS}} \left[\sum_{y \in \text{GIFTS}(x)} \text{V}(y) \right].$$

Nested sums are useful in formalizations of utilitarianism. For instance, there are many persons in a world; each person is divisible into instantaneous person-stages; each stage has some degree of pleasure. Suppose, crudely, that utility is just total pleasure (total hedonic value). The hedonic value of a person is the sum, for all his or her stages, of the pleasure of that stage. The hedonic value of a world is the sum, for every person in the world, of the hedonic value of the person. Thus we have a nested sum.

16. Sequences and Operations on Sequences

Sequence or Series. A *sequence* or *series* is a function from a subset of the natural numbers to a set of things. Recall that the natural numbers are just the counting numbers 0, 1, 2, 3 and so on. Unless the context indicates otherwise, *number* just means *natural number*. The purpose of a sequence is to assign an order to a set of things. Informally, you make a sequence by numbering the things in a set.

Although any subset of the numbers can be used for the domain of a sequence, the subset is usually just some *initial part* of the number line or the whole number line. In other words, the domain of a sequence usually starts from 0 or 1 and runs through the higher numbers without gaps. For example, the sequence of capital letter-types in the Roman alphabet is a function from {1, 2, 3. . . 26} to {A, B, C, . . . Z}.

If S is a sequence from {0, . . . n} to the set of things T, then S(n) is the n-th item in the sequence (it is the n-th item in T as ordered by S). We use a special notation and write S(n) as S_n. We write the sequence as $\{S_0, \ldots S_n\}$ or sometimes as $\{S_n\}$.

Given a sequence $\{S_0, \ldots S_n\}$ of numbers, we can define the sum of its members by adding them in sequential order. We use a variable i to range over the sequence. The notation "for i varying from 0 to n" means that i takes on all the values from 0 to n in order. To say that i varies from 0 to 3 means that i takes on the values 0, 1, 2, 3. The sum, for i varying from 0 to 3, of S_i is $S_0 + S_1 + S_2 + S_3$. Formally, a sequential sum is written like this:

$$\text{the sum, for } i \text{ varying from 0 to } n \text{, of } S_i = \sum_{i=0}^{n} S_i.$$

Given a sequence $\{S_0, \ldots S_n\}$ of sets, we can define the union of its members by unioning them in sequential order. The sequential union looks like this:

the union, for i varying from 0 to n, of $S_i = \bigcup_{i=0}^{n} S_i$.

Analogous remarks hold for sequential intersections:

the intersection, for i varying from 0 to n, of $S_i = \bigcap_{i=0}^{n} S_i$.

17. Cardinality

Cardinality. The *cardinality* of a set is the number of members of the set. The cardinality of a set S is denoted |S| or less frequently as #S. The cardinality of a set is n iff there exists a 1-1 correspondence (a bijection) between S and the set of numbers less than n. The cardinality of the empty set {} is 0 by default.

Let's work out a few examples involving cardinality. The set of numbers less than 1 is {0}. There is a 1-1 correspondence between {A} and {0}. This correspondence just pairs A with 0. So the cardinality of {A} is 1. The set of numbers less than 2 is {0, 1}. There is a correspondence between {A, B} and {0, 1}. The correspondence is A → 0 and B → 1. So the cardinality of {A, B} is 2. The set of numbers less than 3 is {0, 1, 2}. There is a correspondence between {A, B, C} and {0, 1, 2}. This is A → 0 and B → 1 and C → 2. So the cardinality of {A, B, C} is 3.

Equicardinality. Set S *is equicardinal with* set T iff there exists a 1-1 correspondence between S and T. This definition of equicardinality is also known as *Hume's Principle:* "When two numbers are so combined as that the one has always a unit answering to every unit of the other, we pronounce them equal" (Hume, 1990: bk. 1, part 3, sec. 1). Consider the fingers of your hands (thumbs are fingers). You can probably pair them off 1-1 just by touching them together, left thumb to right thumb, left forefinger to right forefinger, and so on. If you can do that, then you've got a 1-1 correspondence that shows that the set of fingers on your left hand is equicardinal with the set of fingers on your right hand. We'll make heavy use of equicardinality in Chapters 7 and 8 on infinity.

Averages. The average of some set of numbers is the sum of the set divided by the cardinality of the set. Hence if S is a set of numbers, then

Average(S) = ΣS / |S|.

The average happiness of a set of people is the sum of their happinesses divided by the number of people in the set. Thus if S is a set of people and H(x) is the happiness of each person x in S, we can define

$$\text{the average happiness of S} = \left[\sum_{x \in S} H(x) \right] / |S|.$$

18. Sets and Classes

It's sometimes said that some collections are too big to be sets. For example, there is no set of all sets. But this isn't really because there are too many sets. It isn't because the set of all sets is too big to be a set. It's because the set of all sets is too general to be a set. There are no constraints on the complexities of the objects that are sets. The collection of all sets is wide open. This openness leads to trouble.

Russell's Paradox. Let R be the set of all sets. If R exists, then we get *Russell's Paradox.* Here's the logic. The axioms of set theory say that for every set x, x is not a member of x. In other words, x is not a member of itself; x excludes itself. Since every set is self-excluding, the set of all sets is just the set of all self-excluding sets:

$$R = \{ x \mid x \text{ is a set} \} = \{ x \mid x \notin x \}.$$

Now what about R itself? Either R is a member of itself or not. Suppose R is not a member of itself, so that R $\notin$ R. But for any x, if $x \notin x$, then $x \in$ R. Hence if R $\notin$ R, then R $\in$ R. The flip side is just as bad. Suppose R is a member of itself, so that R $\in$ R. But for all x, if $x \in$ R, then $x \notin x$. Hence if R $\in$ R, then R $\notin$ R. Putting this all together, R $\in$ R iff R $\notin$ R. And that's absurd. There is no set R.

One way to prevent problems like Russell's Paradox is to place restrictions on the use of predicates to define sets. This is how standard set theory avoids Russell's Paradox. In standard set theory, you can't just form a set with a predicate P. You need another set. You can't write $\{ x \mid x \text{ is P}\}$. You need to specify another set from which the x's are taken. So you have to write $\{ x \in y \mid x \text{ is P} \}$. Given some set y, we can always define $\{ x \in y \mid x \notin x \}$. But that's just y itself. Hence we avoid Russell's Paradox.

For many mathematicians and philosophers, this seems like a defective way to avoid Russell's Paradox. After all, there is some sense in which R is a collection. Namely, the intuitive sense that every property has an *extension*. The extension of a property is the collection of all and only those things that have the property. For instance, the extension of the property is-a-dog is the set $\{ x \mid x \text{ is a dog} \}$. We ought to try to save the existence of R if we can. Of course, we know that it can't be a *set*. The solution is to distinguish collections that are sets from those that are not sets.

Classes. We turn to a more general theory of collections. According to *class theory*, all collections are classes. Any class is either a set or else a *proper class*. Every set is a class, but not all classes are sets. Specifically, the proper classes are not sets. What's the difference? A set is a member of some other class. In other words, if X is a set, then there is some class Y such that X ∈ Y. For example, if X is a set, then X is a member of its power set. That is, X ∈ pow X. But a proper class is not a member of any class. In other words, if X is a proper class, then there is no class Y such that X ∈ Y.

The Class of all Sets. There is a class of sets. We know from Russell's Paradox that the class of all sets can't be a set. It is therefore a proper class. It is not a member of any other class. The *proper class of all sets* is V. That is

$$V = \{ x \mid x \text{ is a set} \}.$$

Of course, V is just the collection R from Russell's Paradox. But now it isn't paradoxical. Since V is a proper class, and *not a set*, V is not a member of V. After all, V only includes *sets*. Class theory thus avoids Russell's Paradox while preserving the intuition that every property has an extension. Class theory is more general than set theory. For every property, the extension of that property is a class. Sometimes, the extension is a set, but in the most general cases, the extension will be a proper class. For example, the extension of the property is-a-finite-number is a set. It is just the set {0, 1, 2, 3, . . .}. But the extension of is-a-set is not a set. It is the proper class V.

A good way to understand the difference between sets and proper classes is to think of it in terms of the hierarchy of sets. A class *is a set* if there is some partial universe of V in which it appears (for partial universes, see Chapter 1, sec. 14). To say that a class appears in some partial universe is to say that there is some *maximum complexity* of its members. A class is a set if there is a cap on the complexity of its members. For example, recall the von Neumann definition of numbers from Chapter 1, sec. 15. If we identify the numbers with their von Neumann sets, then the class {0, 1, 2} appears in the fourth partial universe V_4. Therefore, {0, 1, 2} is a set.

A class *is a proper class* if there is no partial universe in which it appears. One way this can happen is if the class includes an object from every partial universe. For example, the class of sets is a proper class because new sets are defined in every partial universe – there is no partial universe in which *all* sets are defined.

As another example, the class of unit sets is proper. The transition from each partial universe to the next always produces new unit sets. Suppose you say U is the *set* of all unit sets. Since every set has a power set, it follows that U has a power set. But {U} is a member of the power set of U. Since {U} has only one member, it follows that {U} is a member of U. But this is exactly the kind of circularity that is forbidden in set theory. Hence there is no set of unit sets. It was wrong to say that U is a set. On the contrary, U is the *proper class* of unit sets. Since U is a proper class, it is not a member of any more complex class. There is no class {U} such that U is a member of {U}.

To say that there is no partial universe in which a proper class appears is to say that there is no complexity cap on its members. There is no complexity cap on unit sets. For any set x, no matter how complex, there is a more complex unit set $\{x\}$.

3

MACHINES

1. Machines

Machines are used in many branches of philosophy. Note that the term "machine" is a technical term. When we say something is a machine, we mean that it has a certain formal structure; we don't mean that it's made of metal or anything like that. A machine might be made of organic molecules or it might even be made of some immaterial soul-stuff. All we care about is the formal structure. As we talk about machines, we deliberately make heavy use of sets, relations, functions, and so on. Machines are used in metaphysics and philosophy of physics. You can use machines to model physical universes. These models illustrate various philosophical points about space, time, and causality. They also illustrate the concepts of emergence and supervenience. Machines are also used in philosophy of biology. They are used to model living organisms and ecosystems. They are used to study evolution. They are even used in ethics, to model interactions among simple moral agents. But machines are probably most common in philosophy of mind.

A long time ago, thinkers like Hobbes (1651) and La Mettrie (1748) defended the view that human persons are machines. But their conceptions of machines were imprecise. Today our understanding of machines is far more precise. There are many kinds of machines. We'll start with the kind known as *finite deterministic automata.* As far as we can tell, the first philosopher to argue that a human person is a finite deterministic automaton was Arthur Burks in 1973. He argues for the thesis that, "A finite deterministic automaton can perform all natural human functions." Later in the same paper he writes that, "My claim is that, for each of us, there is a finite deterministic automaton that is behaviorally equivalent to us" (1973: 42). Well, maybe there is and maybe there isn't. But before you try to tackle that question, you need to understand finite deterministic automata. And the first thing to understand is that the term *automaton* is somewhat old-fashioned. The more common current term is just *machine*. So we'll talk about machines.

2. Finite State Machines

2.1 Rules for Machines

A *machine* is any object that runs a *program.* A program guides or governs the behavior of its machine. It is a lawful pattern of activity within the machine – it is the *nature* or *essence* of the machine. Suppose that some machine M runs a program P. Any program P is a tuple (I, S, O, F, G). The item I is the set of possible *inputs* to M. The item S is the set of possible *states* of M. The item O is the set of possible *outputs* of M. The item F is a *transition relation* that takes each (input, state) pair onto one or more states in S. It is a relation from $I \times S$ to S. The item G is an *output relation* that takes each (input, state) pair onto one or more outputs in O. It is a relation from $I \times S$ to O.

A machine is *finite* iff its set of program states is finite. Such a machine is also known as – you guessed it – a *finite state machine* (an FSM). A machine is *infinite* iff its set of states is infinite. A machine is *deterministic* iff the relations F and G are functions. For a deterministic machine, the item F is a transition function that maps each (input, state) pair onto a state. In symbols, $F: I \times S \rightarrow S$. And the item G is an output function that maps each (input, state) pair onto an output. In symbols, $G: I \times S \rightarrow O$. A machine is *non-deterministic* iff either F is not a function or G is not a function (they are one-many or many-many). We'll only be talking about deterministic machines.

There are many ways to present a machine. One way is to display its program as a list of *dispositions*. Each disposition is a rule of this form: if the machine gets input *w* while in state *x*, then it changes to state *y* and produces output *z*. For example, consider a simple robot with three emotional states: calm, happy, and angry. One disposition for this emotional robot might look like this: if you get a smile while you're calm, then change to happy and smile back. Of course, the robot may have other dispositions. But it's important to see that the emotional robot has all and only the dispositions that are defined in its program. It has whatever dispositions we give it. We might give it dispositions that allow it to learn – to form new dispositions, and to modify its original programming. But even then, it won't have any undefined dispositions. It is wholly defined by its program.

Another way to present a machine is to display its program as a *state-transition network*. A state-transition network has circles for states and arrows for transitions. Each arrow is labeled with <input / output>. Figure 3.1 shows how a single disposition is displayed in a state-transition network. Figure 3.2 shows some of the state-transition network for the emotional robot.

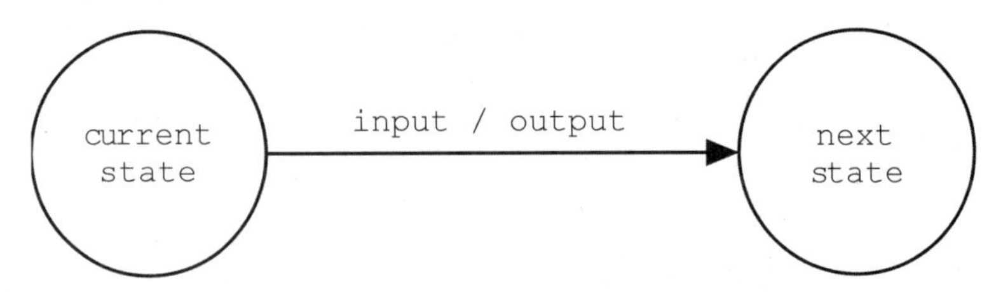

Figure 3.1 Diagram for a single disposition.

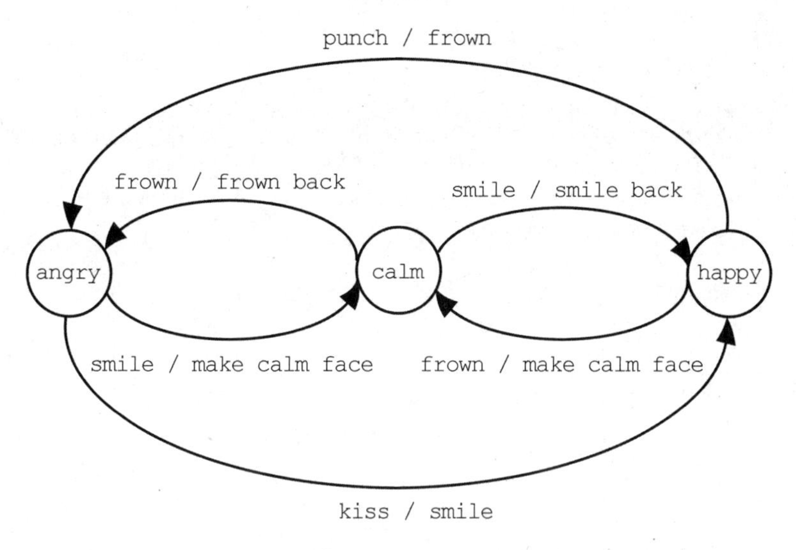

Figure 3.2 State-transition network for an emotional robot.

We can also define the robot by writing its components in set-theoretic notation. So

I = {frown, smile, punch, kiss};
S = {angry, calm, happy};
O = {frown back, smile back, make calm face}.

Table 3.1 details the function F while Table 3.2 details the function G. The functions F and G in these tables are complete – they include all the dispositions of the robot.

(frown, calm)	→	angry
(smile, calm)	→	happy
(punch, calm)	→	angry
(kiss, calm)	→	happy
(frown, happy)	→	calm
(smile, happy)	→	happy
(punch, happy)	→	angry
(kiss, happy)	→	happy
(frown, angry)	→	angry
(smile, angry)	→	calm
(punch, angry)	→	angry
(kiss, angry)	→	happy

Table 3.1 The function F.

(frown, calm)	→	frown back
(smile, calm)	→	smile back
(punch, calm)	→	frown back
(kiss, calm)	→	smile back
(frown, happy)	→	make calm face
(smile, happy)	→	smile back
(punch, happy)	→	frown back
(kiss, happy)	→	smile back
(frown, angry)	→	frown back
(smile, angry)	→	make calm face
(punch, angry)	→	frown back
(kiss, angry)	→	smile back

Table 3.2 The function G.

2.2 The Careers of Machines

Configurations. Suppose machine M runs a program P = (I, S, O, F, G). A *configuration* of M is any quadruple of the form (program P, input *i*, state *s*, output *o*). So the set of all configurations of M is the set of all quadruples of the form (program P, input *i*, state *s*, output *o*). The set of configurations of M is this Cartesian product:

the configurations of M = $C_M = \{P\} \times I \times S \times O$.

Successors. The set of configurations is organized by a successor relation. For any configurations *x* and *y* in C_M, we say

x is a successor of y iff
the program of *x* is the program of *y*; and
the input of *x* is any member of I; and
the state of *x* is the application of F to the input and state pair of *y*; and
the output of *x* is the application of G to the input and state pair of *y*.

It's convenient to use a table to illustrate the successor relation. Since any configuration has four items, the table has four rows: the program; the input; the state; the output. The columns are the configurations. Table 3.3 shows a configuration and one of its successors. The program of the successor is P. The input of the successor is some new input from the environment of the machine. This input has to be in I. We denote it as *i**. The state of the successor is the result of applying the function F to the pair (*i*, *s*). The output of the successor is the result of applying the function G to the pair (*i*, *s*). If a machine has many possible inputs, then any configuration has many possible successors. For such machines, the successor relation is not a function.

	Configuration	**Successor**
Program	P	P
Input	*i*	*i**
State	*s*	F(*i*, *s*)
Output	*o*	G(*i*, *s*)

Table 3.3 A configuration and one of its successors.

Careers. At any moment of its existence, a machine is in some configuration. If a machine persists for a series of moments, then it goes through a series of configurations. Suppose the series of moments is just a series of numbers {0, . . . *n*}. A series of configurations of a machine is a possible *history*, *career* or *biography* of the machine. A career is a function from the set of moments {0, . . . *n*} to the set of configurations C_M. We say

a series H *is a career of* a machine M iff
H_0 is some initial configuration (P, i_0, s_0, o_0); and
for every *n* in the set of moments, H_{n+1} is a successor of H_n.

A table is a good way to display a career. The table has four rows. It has a column for every moment in the career. Each column holds a configuration, labeled with its moment. For example, Table 3.4 illustrates a career with four configurations.

	Moment 0	**Moment 1**	**Moment 2**	**Moment 3**
Program	P	P	P	P
Input	i_0	i_1	i_2	i_3
State	s_0	$s_1 = F(i_0, s_0)$	$s_2 = F(i_1, s_1)$	$s_3 = F(i_2, s_2)$
Output	o_0	$o_1 = G(i_0, s_0)$	$o_2 = G(i_1, s_1)$	$o_3 = G(i_2, s_2)$

Table 3.4 A career made of four configurations.

Looking at Table 3.4, you can see how each next configuration in a career is derived from the previous configuration. Technically, H_{n+1} is (P, i_{n+1}, s_{n+1}, o_{n+1}) where i_{n+1} is any input in I; $s_{n+1} = F(i_n, s_n)$; and $o_{n+1} = G(i_n, s_n)$. Different inputs to a machine determine different careers. Tables 3.5 and 3.6 show two careers for our emotional robot. Note that the input received at any moment has its effect at the next moment. For example, the punch received at moment 1 has its effect at moment 2 – it makes the robot angry then.

	Moment 0	**Moment 1**	**Moment 2**
Program	P	P	P
Input	smile	punch	kiss
State	calm	happy	angry
Output	make calm face	smile back	frown

Table 3.5 A career for the emotional robot.

	Moment 0	**Moment 1**	**Moment 2**
Program	P	P	P
Input	smile	frown	smile
State	calm	happy	calm
Output	make calm face	smile back	make calm face

Table 3.6 Another career for the emotional robot.

Any non-trivial machine (a machine with more than one disposition) has many possible careers. We can formalize the set of careers of a machine M like this:

the careers of M = H_M = { h | h is a career of M }.

For any non-trivial machine, the configurations in the possible careers can be organized into a diagram. The diagram is a connect-the-dots network. Each dot is a configuration and each connection is an arrow that goes from one configuration to a possible successor. So the diagram has the form of a branching tree. Each path in this tree is a possible career of the machine. A tree is a useful way to illustrate branching possibilities or alternatives – the idea is that any non-trivial machine has many possible alternative careers. For a living machine, these are its possible lives. We'll use the idea of many possible lives when we talk about choices in Chapter 6 on utilitarianism.

Figure 3.3 shows a few parts of a few careers of our emotional robot. The robot starts out in a calm state with a calm face. At this initial time, someone is smiling at it. No matter what next input it gets from its environment, it will change to a configuration in which it is happy and smiling back. But since the robot can get many inputs, the initial configuration branches.

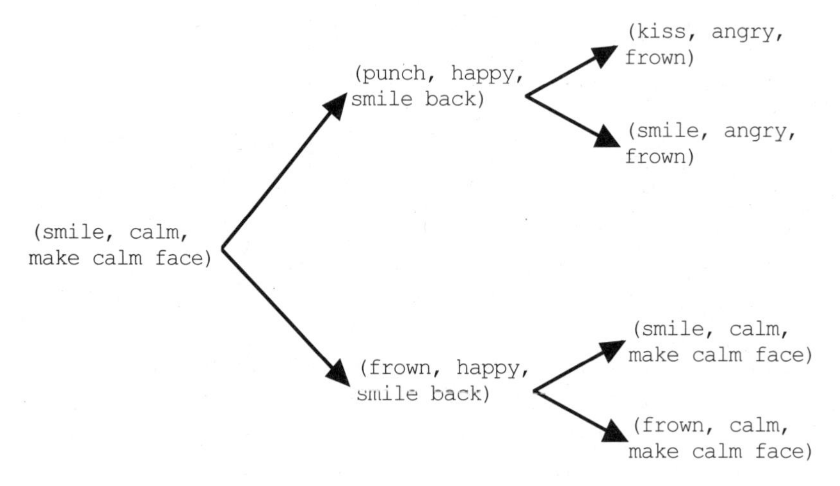

Figure 3.3 Some partial careers for our emotional robot.

2.3 Utilities of States and Careers

A *utility function* for a machine maps each state of the machine onto its happiness. The happiness of a state is just the pleasure that the machine experiences when it is in that state. Clearly, this is an extremely simplified conception of utility. But it serves our purposes. For example, we might assign utilities to our emotional robot like this:

Utility(angry) = -1;
Utility(calm) = 0;
Utility(happy) = 1.

The utility of a configuration is the utility of its state. The utility of a career is the sum of the utilities of its configurations. Suppose the career H of some machine runs from configuration 0 to n. The i-th configuration of that career is H_i. We thus write the utility of a career H as the sum, for i varying from 0 to n, of Utility(H_i). Here it is:

$$\text{UTILITY(H)} = \sum_{i=0}^{n} \text{Utility}(H_i).$$

3. Networks of Machines

3.1 Interacting Machines

A machine can't exist by itself. It can't live alone. A machine has to get its inputs from something and has to give its outputs to something. It has to interact with something else. Suppose a machine X interacts with a partner Y. The outputs from X are inputs to Y. The outputs from Y are inputs to X. So the partner Y also has a set of inputs and outputs. And Y also has some internal state. Thus Y is another machine.

Machines interact with machines. A system of interacting machines is a *network* of machines. The simplest network is a pair of interacting machines. But more complex arrangements are obviously possible. The next more complex network is a triangle of three interacting machines. Figure 3.4 shows three interacting machines. Each circle is a machine. The circles contain the features of the machines. Each arrow stands for an interaction relation. An arrow from machine x to y indicates a signal sent from x to y. This signal is output from machine x and is input to machine y.

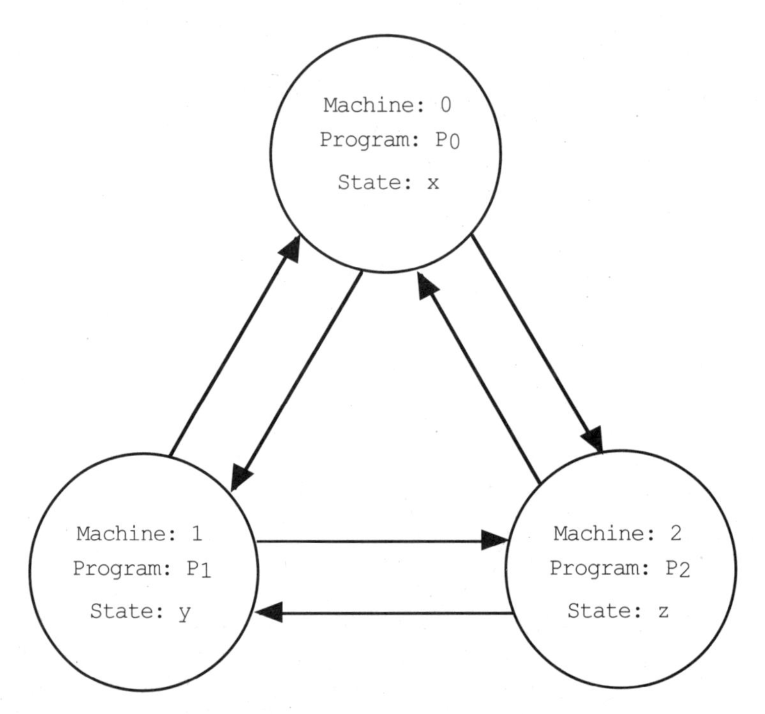

Figure 3.4 A network with three interacting machines.

3.2 Machines in the Network

There are many machines in a network. To distinguish the machines, we give them numbers. For our triangular network, the machines are thus Machine 0; Machine 1; Machine 2. We'll just refer to them as M_0, M_1, and M_2. The first item in any definition of a network of interacting machines is the set of machines in the network:

$$L = \{ M_0, M_1, M_2\}.$$

Any collection of machines runs in time. Hence the definition of a network of machines has to include a set of times. Each time is a number. So the set of times T is a set of numbers running from 0 to n. Formally,

$$T = \{0, \ldots n\}.$$

For any two machines to interact, they have to share a common range of inputs and outputs. They have to speak the same language. The outputs of each have to be the same as the inputs to each. Here's one way of looking at this. Inputs and outputs are *signals*. Every signal has two parts: its content and its strength. The content is the meaning of the signal. The strength of a signal is inversely proportional to the distance between the machines. Stronger signals indicate lesser distance. If machine x is closer to machine y, then the strength of the signal sent from x to y is greater. Signal contents and strengths are just numbers. Thus

E = the set of signal contents;
W = the set of signal strengths.

The set of signals is the set of all (content, strength) pairs. It is the Cartesian product of E with W. In symbols,

$$\text{Signals} = \text{E} \times \text{W}.$$

The input to any machine in a network includes a signal sent from every other machine. For completeness, this input includes a signal sent from the machine to itself. More precisely, the input to any machine is a function from L to the set of signals:

$$\text{I} = \{ f \mid f: \text{L} \rightarrow \text{Signals} \}.$$

Symmetrical reasoning applies to outputs. The output from any machine in a network includes a signal sent to every machine in the network. Hence, the set of outputs is the same as the set of inputs. The set of outputs is

$$\text{O} = \{ f \mid f: \text{L} \rightarrow \text{Signals} \}.$$

At any time, distinct machines can be in distinct states. Thus each machine has to have its own state. But we don't want to define a distinct set of possible states for every distinct machine – that's too much detail. All the machines in a network can take their states from the same set of possible states. Naturally, distinct machines can operate on distinct subsets of the same set of possible states. States are just numbers. Thus

$$\text{S} = \{0, \ldots z\}.$$

Each machine has its own *program*. The program of a machine defines its causal economy – how it responds to possible stimuli. The program of a machine is the nature or essence of the machine. As expected, a program is a tuple (I, S, O, F, G). The transition function F maps each (input, state) pair onto a next state. Thus F: $\text{I} \times \text{S} \rightarrow \text{S}$. The output function G maps each (input, state) pair onto an output. Thus G: $\text{I} \times \text{S} \rightarrow \text{O}$. Although the sets I, S, and O are the same for all machines in a network, that does not imply that the functions F and G are the same for all machines in that network. There are many different functions from $\text{I} \times \text{S} \rightarrow \text{S}$ and from $\text{I} \times \text{S} \rightarrow \text{O}$. Consequently, the functions F and G can vary from machine to machine. Since F and G can vary, the programs of the machines can vary.

3.3 Configurations in the Network

At any time, every machine in a network has a *configuration.* The configuration of any machine in the network is a quadruple of the form (program, input, state, output). We let P be the set of programs run by machines in the network. The total set of configurations for all the machines in the network is

$$C_M = P \times I \times S \times O.$$

An easy way to display the configuration of any machine is to use a *configuration table.* A configuration table illustrates the configuration of some machine *m* at some moment or time *t*. We put the time above the configuration and the machine on the left.

Since configurations are complex, any configuration table is divided in a complex way. It has to display all the information associated with a configuration. Every configuration has four parts: program, input, state, and output. So every configuration table is divided into four rows. The first row displays the program of the machine. The second row displays the input. Since the input is a function, this row contains an entire sub-table. The third row displays the state of the machine. And the fourth row displays the output. Since the output is a function, this row contains a sub-table. Table 3.7 illustrates the configuration of machine M_0 at some time *t*. The input sub-table shows the signals received by M_0 from the various other machines in our triangular network at time *t*. Likewise, the output sub-table shows the signals sent from M_0 to the other machines in the triangular network at time *t*.

	Moment *t*		
M_0	Program	the program of machine M_0	
	Input	M_0	the signal *received* by M_0 from M_0 at time *t*
		M_1	the signal *received* by M_0 from M_1 at time *t*
		M_2	the signal *received* by M_0 from M_2 at time *t*
	State	the state of M_0 at time *t*	
	Output	M_0	the signal *sent* by M_0 to M_0 at time *t*
		M_1	the signal *sent* by M_0 to M_1 at time *t*
		M_2	the signal *sent* by M_0 to M_2 at time *t*

Table 3.7 The configuration of machine M_0 at time *t*.

3.4 The History of a Network of Machines

Any interesting network persists for some sequence of moments – it has a history. The history of a network is a function that associates each (machine, time) pair with some configuration. Technically, the *history function* H has the form

$$H: L \times T \rightarrow C_M.$$

An easy way to display the history of a network of machines is to use a *history table*. The columns are times while the rows are machines. Thus the rows and columns display L × T. Each table cell located at row *m* and column *t* contains the configuration of the machine *m* at time *t*. That is, each cell contains a configuration table. Table 3.8 shows the form of a history table for our triangular network of machines. Table 3.8 shows three moments for this network. Naturally, the network can persist for much longer.

Since we haven't defined the programs for the machines in the triangular network, we don't fill in the details of the history table in Table 3.8. Our purpose is only to illustrate the form of a history table. Filling it in would be way too much detail. But it's worth attending carefully to one aspect of this detail. At any time, each machine in a network sends a signal to every machine and receives a signal from every machine. The signal sent by machine *x* to machine *y* at time *t* is received by machine *y* from *x* at the next moment of time *t*+1. Table 3.8 illustrates the transmission of outputs from M_0 at time 0. Notice how these signals appear as the inputs to other machines at time 1.

Each column in a history table is a complete *snapshot* of the whole network at some time. So long as all the machines in a network are deterministic (and all the machines we've talked about are deterministic), the facts in a snapshot enable you to compute the next snapshot of the network. Each machine gathers its inputs. Each machine looks at its state. It uses its program to compute its next state and output. These are written into the appropriate places in the next snapshot. All this talk about machines looking at their states and using their programs is merely poetic. Snapshots are mathematical structures. Given one, the next one follows mathematically. The facts recorded in any snapshot imply the facts in the next snapshot. Each snapshot has a *successor*. For any network of deterministic machines, any initial snapshot determines a unique series of snapshots. Continuing with our photographic analogy, we say a series of snapshots is a *movie*. Thus a history table is a movie of a network of machines.

You might wonder what happens if the number of machines changes. The number of machines in our definition of a network is fixed. So how can networks handle change? The idea is this: the number of machines is the maximal size of the network. But not every machine has to be active. Not every row and column in a movie needs to be filled in with an active machine. Some rows and columns can be left blank. More precisely, they are filled with *null machines*. Null machines are always in a special null state. We can use 0 for this state. Their input and output sets are the empty set. They don't interact with other machines. When an old machine leaves the network, its row and column go blank. When a new machine enters the network, a blank row and column become meaningful.

	Moment 0			Moment 1			Moment 2		
M_0	Program			Program			Program		
		M_0			M_0	a		M_0	
	Input	M_1		Input	M_1		Input	M_1	
		M_2			M_2			M_2	
	State			State			State		
		M_0	a		M_0			M_0	
	Output	M_1	b	Output	M_1		Output	M_1	
		M_2	c		M_2			M_2	
M_1	Program			Program			Program		
		M_0			M_0	b		M_0	
	Input	M_1		Input	M_1		Input	M_1	
		M_2			M_2			M_2	
	State			State			State		
		M_0			M_0			M_0	
	Output	M_1		Output	M_1		Output	M_1	
		M_2			M_2			M_2	
M_2	Program			Program			Program		
		M_0			M_0	c		M_0	
	Input	M_1		Input	M_1		Input	M_1	
		M_2			M_2			M_2	
	State			State			State		
		M_0			M_0			M_0	
	Output	M_1		Output	M_1		Output	M_1	
		M_2			M_2			M_2	

Table 3.8 A history of a mechanical universe.

3.5 Mechanical Metaphysics

A network of machines is organized by temporal relations. The time indexes of snapshots determine how machines change. Machines have pasts and futures. A network of machines is also organized by spatial relations. The signal strengths determine the distances between machines. A network of machines is organized by causal relations. The inputs to machines cause them to change according to their programs. Their programs are causal laws. Hence a network of machines is a spatio-temporal-causal system. Since every machine gets all its inputs from and sends all its outputs to machines in the network, any network of machines is closed. It is a closed spatio-temporal-causal system. Recall that many philosophers say that a *universe* is a closed spatio-temporal-causal system.

Hence a network of machines is a universe. Specifically, a *mechanical universe* is a tuple (L, T, E, W, I, O, S, P, H).

4. The Game of Life

4.1 A Universe Made from Machines

A computer game known as *the game of life* shows how networks of machines can model physical systems. The game of life was invented by the mathematician John Conway. A good book about the game of life is Poundstone's (1985) *The Recursive Universe.* Many programs are available for playing the game of life on all kinds of computers, and there are many websites you can easily find which will allow you to download these programs for free or to use them online. Dennett advertises the game of life like this:

> every philosophy student should be held responsible for an intimate acquaintance with the Game of Life. It should be considered an essential tool in every thought-experimenter's kit, a prodigiously versatile generator of philosophically important examples and thought experiments of admirable clarity and vividness. (1991: 37)

Any game of life is a spatio-temporal-causal system. It is a trivial mechanical universe. The game of life is played on a 2-dimensional rectangular grid, like a chess board or a piece of graph paper. This rectangular grid is called the *life grid*. The life grid is the *space* for the life universe. The space of the life universe is thus divided into square points.

Space in the game of life is discrete. A point has neighbors on all sides and at all corners, for a total of 8 neighbors. Points in the game of life have minimal but non-zero extension. Since space is 2-dimensional, each point in the game of life has spatial coordinates on an X axis and a Y axis. The life grid is infinite in all spatial directions. So the spatial coordinates are integers (recall that the *integers* include the negative numbers, 0, and the positive numbers). Time is also discrete in the game of life. Time is divided into indivisible *moments* of minimal but non-zero duration. Moments occur one after another like clock ticks; time does not flow continuously in the game of life, but goes in discontinuous jumps or discrete steps from one clock tick to the next. Each point has a temporal coordinate on a T axis. There is an initial moment in the game of life. It has time coordinate 0. So time coordinates start with 0 and run to infinity. Since there are 2 spatial dimensions and 1 temporal dimension, a game of life occurs in a 3-dimensional (3D) space-time.

Each point in the life grid has an energy level. Its energy level is either 0 or 1. You can think of these values as OFF or ON; RESTING or EXCITED; LIVE or DEAD. If P is the set of points in the life grid, then the distribution of energy to these points is a function f from P to {0, 1}. A distribution of some values to points is a *field*. So the function f is an *energy field*. A field is *boolean* iff the numbers are taken from {0, 1}. So the energy field is a boolean field. Boolean values are *binary digits* – they are *bits*. So the energy field in the game of life is also known as a *bit field*. Energy, time, and space are all discrete in the game of life. It is a trivial example of a discrete mechanical system. You play the game of life by setting up the initial values of the energy field (assign a 0 or 1 to each point at the initial instant of time), and then watching the changes in the energy field of the whole grid from moment to moment. For simple examples, you can calculate these changes by hand; but it's more fun to watch them unfold on a computer screen.

4.2 The Causal Law in the Game of Life

As time goes by, the energy field of the whole life grid changes as the energies of individual points change. The energy field evolves according to the basic *causal law* of the game of life. The law is universal: it is the same for all points. It refers only to the past energies (0 or 1) of neighboring points in the grid, so that the present energy field of the grid depends only on the past energy field. The causal law involves only spatial and temporal neighbors. It is a *local* causal law. Suppose that E(p) is the temporal predecessor of any point p and that N(p) is the set of spatial neighbors of p. Each point's present energy field value depends on the sum of the energy field values of the spatial neighbors of its temporal predecessor. To get this sum, we start with point p; we then apply E to p to get its temporal predecessor E(p); we then apply N to E(p) to get the set of spatial neighbors N(E(p)). Finally, we just sum the energy field values of the cells in N(E(p)).

Each point in the game of life is a finite state machine. The machine computes the local causal law. Since the law is the same for all points, each point is a machine M running the same program P = (I, S, O, F, G). The possible inputs to M are the sums of the values of N(E(p)). Since a point has 8 predecessors, these inputs vary from 0 (no predecessors are ON) to 8 (all predecessors are ON). Hence I = {0, . . . 8}. The state of a point is its energy level. Hence the possible states are in S = {0, 1}. The output of a point is a quantum of energy sent to the future. The possible outputs are in O = {0, 1}.

The functions F and G determine how the energies and outputs of points change. The energy of a point is 0 or 1 at the next moment depending on two factors: (1) its current energy; and (2) the total number of past neighbors that are 1. If a point is 1 at this moment, then it stays 1 in the next moment if it has either 2 or 3 neighbors that are 1; otherwise it changes to 0. If a point is 0 at this moment, then it changes to 1 in the next moment if it is surrounded by exactly 3 neighbors that are 1; otherwise it stays 0. The functions F and G are shown as a state-transition network in Figure 3.5. They are also displayed in Table 3.9. Given any (input, state) pair, the columns F and G show the values of the functions.

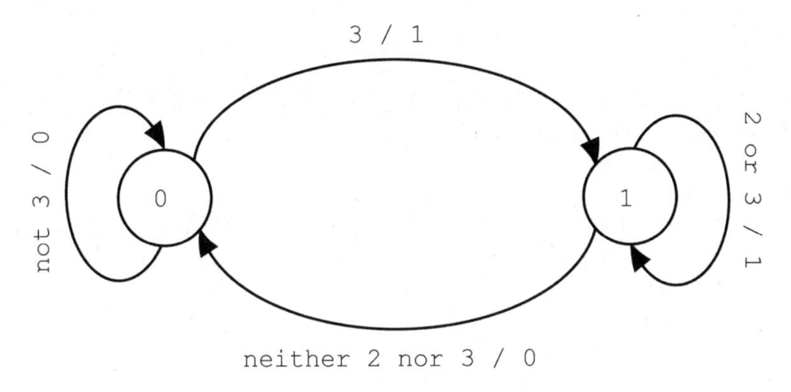

Figure 3.5 State-transition network for the game of life.

Input	State	F	G
0	0	0	0
1	0	0	0
2	0	0	0
3	0	1	1
4	0	0	0
5	0	0	0
6	0	0	0
7	0	0	0
8	0	0	0
0	1	0	0
1	1	0	0
2	1	1	1
3	1	1	1
4	1	0	0
5	1	0	0
6	1	0	0
7	1	0	0
8	1	0	0

Table 3.9 The state-transition table for the game of life.

Cellular Automaton. A cellular automaton is any system in which space and time are discrete; all physical properties are discrete; and a single causal rule is operative at each point in space-time (Toffoli & Margolus, 1987). The points in a cellular automaton are also called, not surprisingly, *cells*. The game of life is obviously a cellular automaton. But there are many others besides the game of life. Cellular automata are used extensively in artificial life, hence in the philosophy of biology. And they play increasingly important roles in all sciences. Some physicists argue that our universe is a cellular automaton (e.g.,

Fredkin, 1991, 2003). Of course, that's highly controversial. But anyone interested in the logical foundations of physics should study cellular automata.

4.3 Regularities in the Causal Flow

Figure 3.6 illustrates how the causal law acts on a horizontal bar of three ON cells. The law changes the horizontal bar into a vertical bar. The law then changes that vertical bar back into a horizontal bar. The oscillating bar of three ON cells is known as the *blinker*. Figure 3.7 shows a pattern that moves. This mobile pattern is known as the *glider*. Although it looks like the cells in the glider are moving, they are not. The cells stay put. The motion of the glider is the motion of a pattern of cell values. It is like the motion of a pattern of light bulb illuminations on a scoreboard. The pattern of illumination moves although the light bulbs stay put. You can construct many different kinds of patterns on the life grid. If you're familiar with the notion of supervenience, you might consider the idea that these patterns *supervene* on the life grid.

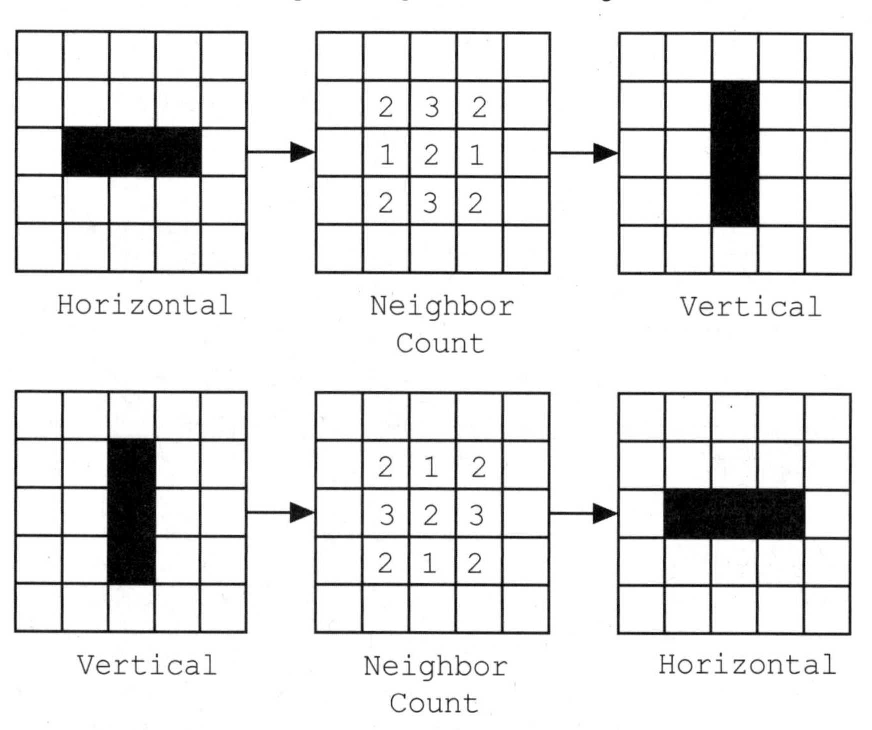

Figure 3.6 Transformations of patterns on the life grid.

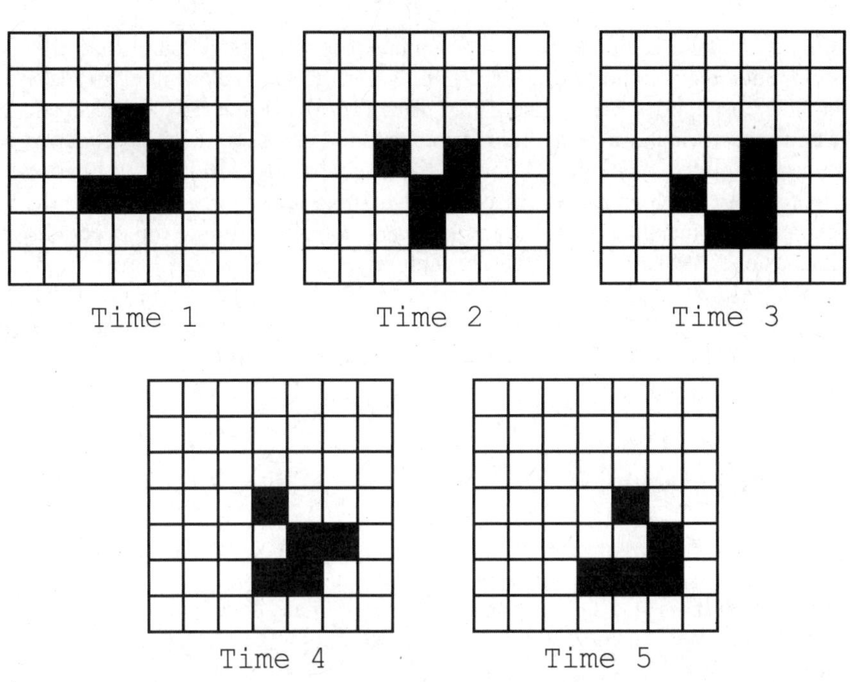

Figure 3.7 The motion of the glider.

We've been talking about blinkers and gliders as if they are things that exist in their own right. But what are they? One way to individuate them and to treat them as particulars is to identify them with sets of points. For example, the blinker is a certain set of all energized points. Another way is to think of each of these things as its own independent energy field. Such an energy field is a *presence function*. More precisely, the presence function for a particular distribution of energies to points is a function from the points in the life grid to {0, 1}. The presence function is 1 if the point is energized in the pattern and it is 0 otherwise. For the blinker, the presence function is 1 for all the energized points in the blinker, and is 0 for all other points on the life grid.

We've mentioned simple patterns like blinkers and gliders, but these are not the only patterns in the game of life. The game of life is famous for generating an enormous variety of patterns. Poundstone (1985) provides an impressive analysis of these patterns. Many large catalogs of patterns in the game of life are available on the Internet. Static patterns are known as *still lifes*. For example, a *block* is just 4 ON cells arranged in a square. It just stays in the same shape in the same place. *Oscillators* are patterns that don't move but that do change their shapes (so the blinker is an oscillator). The glider is an example of a moving pattern. Such patterns are also known as *spaceships* or *fish*. The glider is not the only mover – there are movers of great complexity and beauty.

Simpler patterns can be assembled to make more complex patterns. For instance, four blinkers can be assembled to make a *stoplight*. Two blocks and a *shuttle* can be assembled to make a complex oscillator (see Poundstone, 1985: 88). Four blocks and two *B heptomino shuttles* can be assembled to make a complex oscillator (Poundstone, 1985: 89). Two shuttles and two blocks can be assembled to make a *glider gun* (Poundstone, 1985: 106). The glider gun generates gliders as output. Many of these patterns are machines in their own right – they are higher level machines that supervene on the life grid. Many higher level machines are known (guns, rakes, puffer trains, breeders – see Poundstone, 1985: ch. 6). If you're familiar with the idea of a logic gate, you might be interested to know that logic gates can be constructed in the game of life (see Poundstone, 1985: ch. 12). It is possible to make a self-reproducing machine that behaves like a simple living organism (see Poundstone, 1985: ch. 12).

5. Turing Machines

A naked finite state machine (FSM) is limited in many ways. For example, if an FSM is going to carry any information about its past, it has to carry that information in its states. But it has only finitely many states. So its memory is only finite. We can remedy this defect by equipping an FSM with an external memory.

A simple way to do this is to let the memory be a tape. The tape is a thin strip of material – perhaps a strip of paper. Since we're not concerned with the physical details, the tape can be as long as we want. We define it to be endlessly long in both directions. The tape is divided into square cells. Each cell can store a symbol from an alphabet. The alphabet can be big or it can be as small as just two symbols (e.g., blank and 1).

The FSM is always positioned over some cell on the tape. It can read the symbol that is in that cell. The tape thus provides the FSM with its input. The FSM can also write to the tape. It does this by erasing any contents of the cell and printing a new symbol in the cell. The tape thus gets the output of the FSM. Finally, the FSM can move back and forth over the tape. It can move one cell to the right or to the left. It can also stay put.

An FSM plus a tape is known as a *Turing Machine*. Such machines were first described by the British mathematician Alan Turing (1936). Since Turing Machines involve infinitely long tapes, they are infinitely complex. Assuming that there are no infinitely complex physical structures in our universe, there are no actual physical Turing Machines. A Turing Machine is an ideal or merely possible device. It is like a perfectly smooth or frictionless plane in physics. Turing Machines are sometimes used in the philosophy of mind. They are essential for any philosophical understanding of the notion of computation. Much has been written about Turing Machines, and we don't intend to repeat it. We're only talking very briefly about Turing Machines for the sake of completeness. To learn more about Turing Machines, we recommend Boolos & Jeffrey's excellent (1989).

A Turing Machine (TM) has two parts: its FSM and its tape. The FSM is sometimes called the TM's *controller* or *head*. The FSM has a finite number of internal states. It also has a finite number of possible inputs and outputs. These are the symbols it can read from and write to the tape. Hence its input set = its output set = the alphabet. A TM has a finite set of dispositions. Each disposition is a rule of this form: if the machine is in state *w* and it reads a symbol *x*, then it performs an action *y* and changes to state *z*. An action is either (1) moving one cell to the left on the tape; (2) moving one cell to the right on the tape; or (3) writing some symbol on the tape. One of the states is a special *halting* state. When the TM enters that state, it stops.

We illustrate TMs with a very simple example. The alphabet in our example is the two symbols, blank and 1. We use # to symbolize the blank. Our example starts with a tape on which a single 1 is written. The head starts either on top of this single 1 or to the left of it. If it is right on top of that 1, then it halts. If it is to the left, it fills in the tape with 1s, always moving right, until it encounters the 1 that was already written. Then it halts.

The machine has 3 states: Start, Moving, and Check. It has two inputs, # and 1. Since it has a disposition (a rule) for each state-input combination, it has six rules. The machine table is Table 3.10. Notice that the machine should never perform the third rule – if it's in the Moving state, there should always be a 1 underneath the head. But for completeness, every state-input combination should have a rule. If the machine is in the Moving state and there's a blank under the head, something has gone wrong. So the machine just halts. We note that it's in an erroneous situation. The series of pictures in Figure 3.8 illustrate the operation of the TM on a sample tape. Only a few cells on the tape are shown. The triangle over a cell indicates that the head is positioned over that cell. The state of the head is written above the triangle. For instance, in the first picture, the head is over a cell with a blank (#), and it is in the Start state. The machine goes through 8 steps and halts.

Our sample TM is extremely simple. But don't be fooled. Some TMs are programmable. The program for the machine is stored on the tape, along with the input to that program. So one TM can simulate other TMs. You could program a TM to simulate our simple TM. Some TMs can be programmed to simulate *any* other TM. They are *universal Turing Machines* (UTMs). Anything that any finitely complex computer can do, can be done by a UTM. The computers we use in everyday life – our PCs – are just finitely complex versions of UTMs. More precisely, the computers we use in everyday life are known as von Neumann Machines; von Neumann Machines are just UTMs with finite memory.

An interesting note is that you can make a UTM in the game of life (Rendell, 2002). This brings us to an intriguing line of speculation. Bostrom (2003) develops the idea that you and I might be living in a computer simulation. Since many philosophers have argued that the mind is just a Turing Machine, our minds might be just Turing Machine patterns running in some game of life. What do you think?

Current State	Input	Action	New State
Start	#	print 1	Moving
Start	1	print 1	Halt
Moving	#	print 1	Halt (error)
Moving	1	move right	Check
Check	#	print 1	Moving
Check	1	print 1	Halt

Table 3.10 The machine table for the simple TM.

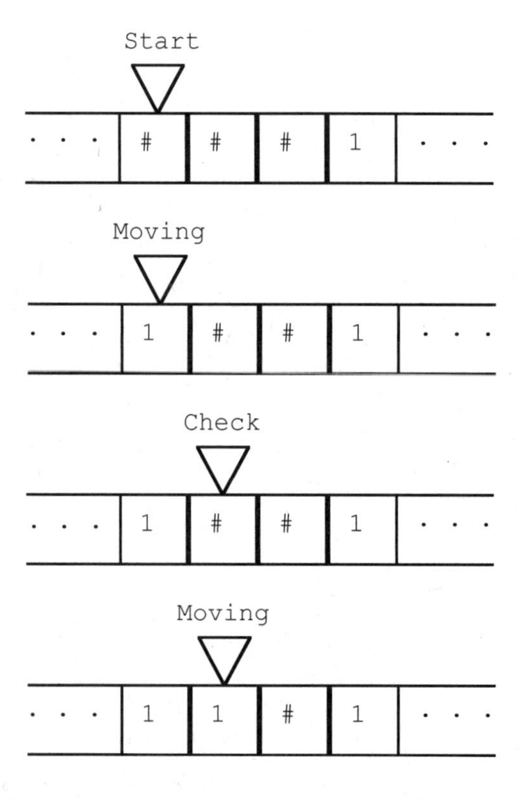

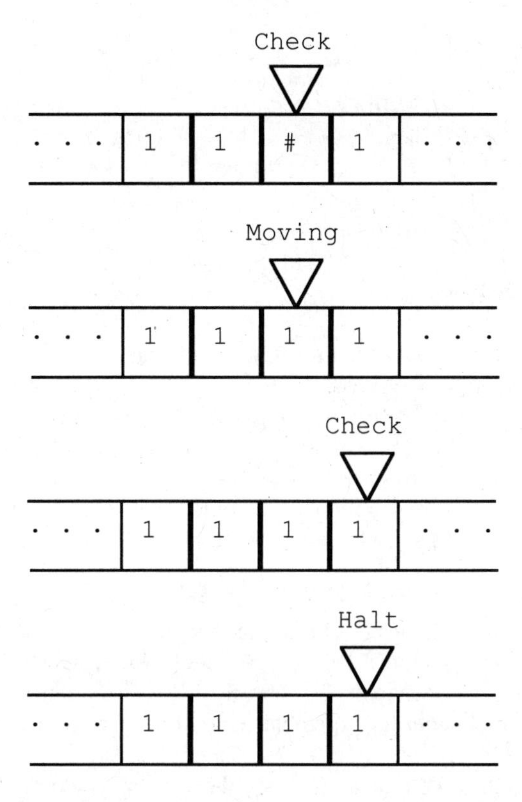

Figure 3.8 A series of snapshots of the simple TM.

4

SEMANTICS

1. Extensional Semantics

1.1 Words and Referents

A *language* is a complex structure with many parts. One of the parts of a language is its *vocabulary*. A vocabulary is a list of words or terms. To keep things simple, we include certain idiomatic phrases among the words. For example, we treat a phrase like "is the father of" as a single word. We might think of a vocabulary as just a set of words. But we want our vocabulary to have some order. There are two options: we could think of it as either a series of words or as an ordered tuple of words. We'll model a vocabulary as an ordered tuple. Thus a vocabulary is (w_0, w_1, . . . w_n) where each w_i is a word.

Some words are proper nouns (we'll just say they're *names*). A name *refers* to a thing. For example, "Socrates" refers to Socrates. Many names can refer to one thing. For example, "Aristocles" refers to Aristocles. And, as it turns out, "Plato" is just a nickname for Aristocles. So "Plato" also refers to Aristocles. Sometimes phrases act like names. For example, "the sun" is a phrase that refers to a single star. Formal semantics – the kind we're doing here – says that every word is a kind of name. Every word refers to some object. Note that the object may be an individual or a set.

Reference Function. Just as a language has a vocabulary, it also has a *reference function*. The reference function maps each word in the vocabulary onto the object to which it refers. It maps the word onto its referent. A language can have many reference functions. Every competent user of some language has his or her own local reference function encoded in his or her brain. In any language community, these local functions are very similar (if they weren't, the members of that community couldn't communicate). For the sake of simplicity, we'll assume that all language users agree entirely on their reference functions. So there is only one.

Names. Every name (every proper noun) in a vocabulary refers to some individual thing. For example, the American writer Samuel Clemens used the pen name "Mark Twain". So the reference function *f* maps the name "Mark Twain" onto the person Samuel Clemens. We can display this several ways:

"Mark Twain" refers to Samuel Clemens;

the referent of "Mark Twain" = Samuel Clemens;

f("Mark Twain") = Samuel Clemens;

"Mark Twain" → Samuel Clemens.

Nouns. A common noun refers to some one thing shared in common by all the things named by that noun. You point to Rover and say "dog"; you point to Fido and say "dog"; so "dog" refers to what they have in common. Generally, "dog" refers to what all dogs have in common. Of course, you could point to arbitrary things and repeat the same name; but that name wouldn't be useful for communication – there could be no agreement about what it means, since the things it refers to have nothing in common. It would be a nonsense name. One hypothesis about commonality is that what things of the same type share in common is membership in a single set. For example, what all dogs share in common is that every dog is a member of the set of dogs.

Assuming that what all Ns have in common is membership in the set of things named by N, we can let the common noun N refer to that set. The noun N is a name that refers to the set of all Ns. Thus f maps "man" onto the set of all men. Figure 4.1 illustrates this with a few men. The set of all things that are named by N is the *extension* of N. So every common noun refers to its extension. We'll use ALL CAPS to designate types of words. For example, NOUN is any common noun. Thus

the referent of NOUN = the set containing each x such that x is a NOUN;

the referent of "man" = the set containing each x such that x is a man;

f("man") = $\{ x \mid x \text{ is a man} \}$.

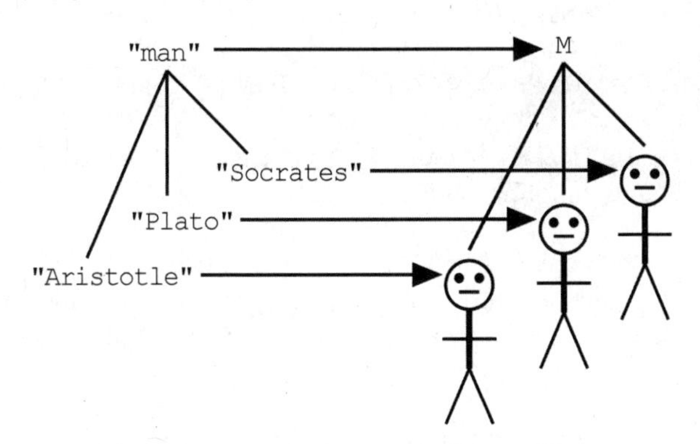

Figure 4.1 Names refer to things; nouns refer to sets of things.

Adjectives. Although adjectives and common nouns have different grammatical roles, they serve the same logical function. An adjective, like a common noun, refers to something shared by all the things named by that adjective. For example, "red" refers to what all red things have in common. An adjective refers to the set of all things that are truly described by the adjective. The reference function f maps each ADJ onto a set (the extension of the ADJ). Here's an example:

the referent of ADJ = the set of all x such that x is ADJ;

the referent of "red" = the set of all x such that x is red;

f("red") = { x | x is red }.

Verbs. Suppose Bob loves Sue and Jim loves Sally. If that's right, then the pairs (Bob, Sue) and (Jim, Sally) have something in common, namely, love. We can use this idea to define extensions for verbs. Verbs are relational terms. The extension of a verb is the set of all tuples of things that stand in that relation. For example, we can pair off men and women into married couples. One example of a married couple is (Eric, Kathleen). Thus (Eric, Kathleen) is in the extension of "is married to". Order is important in relations. The relation "is the child of" has the opposite order of "is a parent of". Pairing off children with parents is different from pairing off parents with children. If Eric is a son of Dean, then (Eric, Dean) is an example of the relation "is a child of". But (Dean, Eric) is not an example of that relation – order matters. Order is important for verbs too: if (Maggie, Emma) is an example of the verb "hits", then it's Emma who gets hit. Example:

the referent of "loves" = { (Bob, Sue), (Sue, Bob),
(Jim, Sally), (Sally, Ray), . . .}.

So Bob & Sue are probably pretty happy – they love each other. But poor Jim is frustrated. Jim loves Sally, but Sally loves Ray. An example with numbers:

the referent of "weighs" = { (Ray, 160), (Jim, 98), (Emma, 80), . . .}.

So Ray weighs 160 pounds while Jim only weighs 98 pounds. Little Emma is 80 pounds, about right for her age.

Verbs need not be active. They are phrases that denote relations. So "is the husband of", "is the wife of", and "is the parent of" are all verbs. As an example, consider the following (partial) extension of the "is the wife of" relation:

"is the wife of" → { (Hillary, Bill), (Barbara, George)}.

This relation is diagrammed in Figure 4.2. In Figure 4.2, each black dot refers to a set and each arrow to an instance of the membership relation. So, the black dot above Hillary is the unit set {Hillary}. The arrows from Hillary and Bill converge on a black dot, the set {Hillary, Bill}. And those two sets are members of the set {{Hillary}, {Hillary, Bill}}. As you'll recall from Chapter 1, sec. 17, this is the ordered pair (Hillary, Bill). Analogous remarks hold for the sets involving Barbara and George.

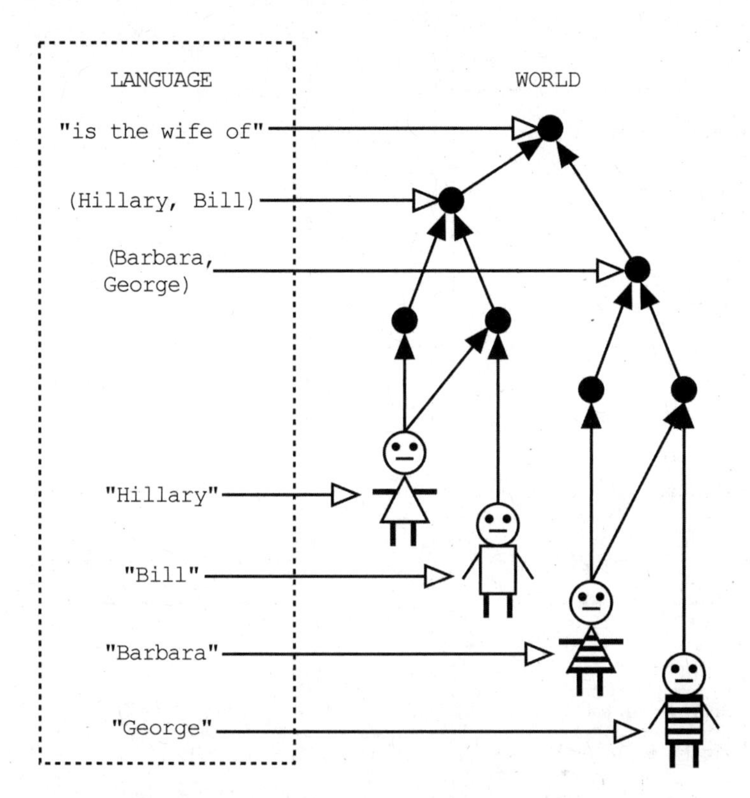

Figure 4.2 A verb refers to a set of ordered pairs.

1.2 A Sample Vocabulary and Model

Model. A language has a vocabulary and a reference function. A *model* for the language is an image of its vocabulary under its reference function. We can display a model by listing each term in a vocabulary on the left side and its referent on the right side. Table 4.1 shows a model for our simple language. The entire "Term" column (regarded as a single ordered list) is the vocabulary. The entire "Referent" column is a model of the language. More formally, suppose our vocabulary is (w_0, w_1, . . . w_n) where each w_i is a word. The image of the vocabulary under the reference function f is ($f(w_0)$, $f(w_1)$, . . . $f(w_n)$).

Term	Referent
"Alan"	A
"Betty"	B
"Charles"	C
"Doreen"	D
"Anny"	a
"Bobby"	b
"Chrissy"	c
"Danny"	d
"person"	{A, B, C, D, a, b, c, d}
"male"	{A, C, b, d}
"female"	{B, D, a, c}
"adult"	{A, B, C, D}
"child"	{a, b, c, d}
"man"	{A, C}
"woman"	{B, D}
"tall"	{A, D, b, c}
"short"	{B, C, a, d}
"is married to"	{(A, B), (B, A), (C, D), (D, C)}
"is the parent of"	{ (A, a), (B, a), (A, b), (B, b), (A, c), (B, c), (C, d), (D, d)}

Table 4.1 A sample reference function.

1.3 Sentences and Truth-Conditions

Since we have discussed names, nouns, adjectives, and verbs, we now have enough terms to make sentences. Sentences are either true or false. The truth-value of a sentence depends on (1) the referents of its words and (2) the syntactic form of the sentence. Each syntactic form is associated with a *truth-condition* that spells out the conditions in which the sentence is true (if those conditions aren't satisfied, then the sentence is false).

A syntactic form lists types of words in a certain order. We write word-types in capital letters. For example: NAME is any name; NOUN is any noun; ADJ is any adjective; VERB is any verb. So we get a syntactic form: <NAME is ADJ>. Note that we enclose syntactic forms in angle brackets. You fill in the form by replacing each word-type with an example of that type. And once you fill in each word-type, the angle brackets change to quotes. Examples of <NAME is ADJ> include sentences like "Socrates is male" and "Sally is happy". For the sake of simplicity, we ignore the little words "a" and "an". Hence the sentence "Socrates is a man" has the form <NAME is NOUN>.

A sentence of the form <NAME is ADJ> is true iff the referent of NAME is a member of the referent of ADJ. Since there can be many reference functions, we need to specify the one we're using. So we need to say that a sentence is true *given* a reference function f. For logical precision, it's good to write out truth-conditions in logical notation. Recall that "the referent of x" is $f(x)$ and "is a member of" is $\in$. So:

<NAME is ADJ> is true given f iff
    $f(\text{NAME}) \in f(\text{ADJ})$.

Thus the truth-value of <NAME is ADJ> is equivalent to the truth-value of $f(\text{NAME}) \in f(\text{ADJ})$. We can also express this by writing

the value of <NAME is ADJ> given f
    $= f(\text{NAME}) \in f(\text{ADJ})$.

The truth-condition for this sentence form is illustrated below:

the value of "Bobby is male" given f
    $= f(\text{"Bobby"}) \in f(\text{"male"})$
    $= \text{b} \in \{\text{A, C, b, d}\}$
    = true;

while

the value of "Bobby is female" given f
    $= f(\text{"Bobby"}) \in f(\text{"female"})$
    $= \text{b} \in \{\text{B, D, a, c}\}$
    = false.

Common nouns behave just like adjectives. So:

the value of <NAME is NOUN> given f
    $= f(\text{NAME}) \in f(\text{NOUN})$.

The truth-condition for this sentence form is illustrated here:

the value of "Bobby is a child" given f
    $= f(\text{"Bobby"}) \in f(\text{"child"})$
    $= \text{b} \in \{\text{a, b, c, d}\}$
    = true;

while

> the value of "Bobby is an adult" given f
> $= f(\text{"Bobby"}) \in f(\text{"adult"})$
> $= \text{b} \in \{\text{A, B, C, D}\}$
> = false.

The truth-condition for a sentence of the form <NAME$_1$ VERB NAME$_2$> looks like this:

> the value of <NAME$_1$ VERB NAME$_2$> given f
> = (the pair formed by $f(\text{NAME}_1)$ and $f(\text{NAME}_2)$) $\in f(\text{VERB})$
> $= (f(\text{NAME}_1), f(\text{NAME}_2)) \in f(\text{VERB})$.

Two examples of this truth-condition are given below:

> the value of <Allan is a parent of Bobby> given f
> $= (f(\text{"Allan"}), f(\text{"Bobby"})) \in f(\text{"is a parent of"})$
> $= (\text{A, b}) \in \{(\text{A, a}), (\text{B, a}), (\text{A, b}), (\text{B, b}), (\text{A, c}), (\text{B, c}), (\text{C, d}), (\text{D, d})\}$
> = true;

> the value of <Allan is married to Doreen> given f
> $= (f(\text{"Allan"}), f(\text{"Doreen"})) \in f(\text{"is married to"})$
> $= (\text{A, D}) \in \{(\text{A, B}), (\text{B, A}), (\text{C, D}), (\text{D, C}) \}$
> = false.

Let's work out some truth-conditions that are more complex. Consider a sentence like "A tall woman is married to a short man". The sentence has this syntactic form:

> <ADJ$_1$ NOUN$_1$ VERB ADJ$_2$ NOUN$_2$>

The sentence "A tall woman is married to a short man" is true iff there exists some x such that x is tall and x is a woman, and there exists some y such that y is short and y is a man, and x is married to y. Let's put this in logical form:

> "A tall woman is married to a short man" is true given f iff
> there exists x such that $x \in f(\text{"tall"})$ & $x \in f(\text{"woman"})$ &
> there exists y such that $y \in f(\text{"short"})$ & $y \in f(\text{"man"})$ &
> $(x, y) \in f(\text{"is married to"})$.

More generally and more formally:

<ADJ$_1$ NOUN$_1$ VERB ADJ$_2$ NOUN$_2$> is true given f iff
(there exists x)(($x \in f(\text{ADJ}_1)$ & $x \in f(\text{NOUN}_1)$) &
(there exists y)(($y \in f(\text{ADJ}_2)$ & $y \in f(\text{NOUN}_2)$) &
$(x, y) \in f(\text{VERB})$)).

The sentence "A tall woman is married to a short man" is in fact true given f since Doreen is a tall woman, Charles is a short man and Doreen is married to Charles.

Consider a sentence like "Every man is male". It has the form

<every NOUN is ADJ>.

The sentence "Every man is male" is true given f iff for every x, if x is a man, then x is male. More set-theoretically: for every x, if x is in the set f("man"), then x is in the set f("male"). Generally:

<every NOUN is ADJ> is true given f iff
(for every x)(if $x \in f(\text{NOUN})$, then $x \in f(\text{ADJ})$).

We know that set X *is a subset of* set Y iff for every x, if x is in X, then x is in Y. Example: f("man") is a subset of f("male") since for every x, if x is a man, then x is male. So we can re-write our last truth-condition in terms of subsets:

<every NOUN is ADJ> is true given f iff
$f(\text{NOUN}) \subseteq f(\text{ADJ})$.

Consider a sentence like "Every woman is a person". It has the form

<every NOUN$_1$ is NOUN$_2$>.

For example, the sentence "Every woman is a person" is true given f iff for every x, if x is a woman, then x is a person. Generally and formally:

<every NOUN$_1$ is NOUN$_2$> is true given f iff
$f(\text{NOUN}_1) \subseteq f(\text{NOUN}_2)$.

2. Simple Modal Semantics

2.1 Possible Worlds

A long tradition in philosophy distinguishes between two modes of existence: actual existence and possible existence. Many philosophers believe that the concept of possible existence has a much larger extension than the concept of actual existence. For example, there are no actual unicorns, but surely there are possible unicorns. So the range of possible things includes some things that are not actual, namely, unicorns. Hence the word "unicorn" refers to a set of possible things. Of course, the concept of the possible includes the concept of the actual. If something is actual, then it is possible.

A semantic theory that recognizes these two modes is a *modal semantic theory.* Modal semantics is also known as *possible worlds semantics,* since it uses possible worlds. There are many ways to do modal semantics. The issues involved are subtle and often highly technical. We're not going to go into them. Nor are we going to develop a full modal theory. We're just going to do enough to illustrate the use of set theory and other formal tools in modal semantics. Our approach is highly simplified.

One way to do possible worlds semantics says that reality is a modal structure (V, W, I, f). The item V is a vocabulary. Item W is a set of possible worlds. What are possible worlds? They are just objects that go into truth-conditions. A great deal has been said about the nature of possible worlds (for an excellent introduction, see Loux 2002: ch. 5). We don't need to go into it here. The item I is a set of individuals. The individuals in I are all possible individuals. These are *possibilia.* They include you and me, Socrates and Plato, Santa Claus and Sherlock Holmes, Bigfoot and the Loch Ness monster, Charles the Unicorn, Lucky the Leprechaun, you name it. Indeed, for any consistent definition of an individual thing, we'd say there's an individual in I that satisfies that definition. Note that all actual things are also possible things. You're actual; therefore, you're possible. But not all possible things are actual. For example, Sherlock Holmes is a non-actual possible thing. Sherlock Holmes is *merely* possible. Finally, the item f is the reference function. It associates each (word, world) pair with the referent of the word at the world. There are other ways to define the reference function. But for the moment, this is the simplest.

At least part of the motivation for possible worlds semantics is that things might have been different. For example, consider the 2000 US Presidential Election. Al Gore did not actually win, but he might have won. What does this mean? It means that Al Gore did not win in our actual world, but he might have won in some other possible world. At our actual world, the name "Al Gore" refers to a man who lost; at some other world, the name "Al Gore" refers to a man who won. More formally,

f("Al Gore", our actual world) = a man who lost;

f("Al Gore", some other possible world) = a man who won.

For simple modal semantics, we assume that the Al Gore who lost is identical to the Al Gore who won. That is, Al Gore lives in many possible worlds. Suppose our world is w_0 and w_1 is a world in which Al Gore won. Specifically,

f("Al Gore", w_0) = Al Gore;

f("Al Gore", w_1) = Al Gore.

More generally, the referent of a proper name does not vary from world to world. A term with such an invariant reference is a *rigid designator*. Proper names are rigid designators. Formally, for any name n, and for any two worlds u and v, $f(n, u) = f(n, v)$.

But if there is only one Al Gore, and he wins in one world but loses in another, then it seems like he both wins and loses. After all, there is only one Al Gore to do both the winning and the losing, even if they are done in different worlds. This appears to be a contradiction. We resolve the contradiction by defining winning and losing relative to worlds. They are *world-indexed properties*. Al Gore wins-in-world-w_1 and he loses-in-world-w_0. But these are distinct properties. Hence winning in one world does not forbid losing in another world. To say that the properties are distinct is to say that they have different extensions in different worlds. More specifically, if "wins" denotes the people who win US Presidential Elections, then

f("wins", w_0) = {Washington, . . . Lincoln, . . . Clinton, GWB, . . .};
f("wins", w_1) = {Washington, . . . Lincoln, . . . Clinton, Gore, . . .}.

Although the reference of names does not vary from world to world, the reference of a *predicate* does. A predicate is a word that is not a proper noun – thus common nouns, adjectives, and verbs are predicates. A given predicate P can have one extension at one world and a different extension at another world. The differences between predicates, from world to world, allow different things to happen at different worlds.

Since f is a function, it has to associate every (word, world) pair with an object. But worlds differ in their contents. For example, Sherlock Holmes does not exist in our world; he exists in some other world(s). So how should we define f for the pair ("Sherlock Holmes", our world)? We can't leave f undefined. The solution is simple. We can just let f map ("Sherlock Holmes", our world) onto Sherlock Holmes. After all, the truth-value at our world of any sentence involving Sherlock depends on the extensions that are defined at our world for the predicates in that sentence. On the one hand, we don't want the sentence "Sherlock is a man" to be true at our world (although we do want it to be true at some other world or worlds). So we don't include him in the extension of "man" at our world (and we do include him in the extension of "man" at some other world or worlds). On the other hand, we do want the sentence "Sherlock is fictional" to be true at our world (although he's not fictional at any of the worlds in which he exists). So we do include him in the extension of "fictional" in our world (and we do not include him in the extension of "fictional" at those worlds in which he exists). Of course, all this raises interesting questions about what it means to be a fictional character. What are your questions about fictional characters? How would you answer those questions?

2.2 A Sample Modal Structure

For the sake of illustration, we discuss a single modal structure (V, W, I, *f*). The vocabulary V is the following ordered tuple of words:

("Allan", "Betty", "Charlie", "Diane", "Eric",
"human", "man", "woman", "loves", "happy", "sad").

The set of worlds W is { w_1, w_2, w_3, w_4}.

The set of individuals I is {A, B, C, D, E}.

The reference function *f* is given in Table 4.2. Words label rows; worlds label columns. Note that names are rigid designators; that is, for any name *n*, and for any two worlds *u* and *v*, $f(n, u) = f(n, v)$. Note that mere reference is not existence. The fact that "Diane" refers to D at w_3 does not imply that Diane exists in w_3; in fact, she does not.

Term	**Reference at Various Worlds**			
	w_1	w_2	w_3	w_4
"Allan"	A	A	A	A
"Betty"	B	B	B	B
"Charlie"	C	C	C	C
"Diane"	D	D	D	D
"Eric"	E	E	E	E
"thing"	{A, B, C, D}	{A, B, C, D, E}	{A, B, C}	{A, B, C, D}
"human"	{A, B, C, D}	{A, B, C, D, E}	{A, B, C}	{A, B, C, D}
"man"	{A, C}	{A, C, E}	{A, C}	{A, C}
"woman"	{B, D}	{B, D}	{B}	{B, D}
"loves"	{(A, B), (B, A), (C, D)}	{(A, B), (B, A), (C, D), (D, E), (E, D)}	{(A, B), (B, A), (C, B)}	{(A, D), (D, A), (C, B)}
"happy"	{A, B, D}	{A, B, D, E}	{A, B}	{A, D}
"sad"	{C}	{C}	{C}	{C, B}

Table 4.2 The sample reference function.

2.3 Sentences and Truth at Possible Worlds

Sentences are true or false at worlds. The truth-value of a sentence can vary from world to world. For example, recall that in our example involving Al Gore, the sentence "Al Gore won" is false at the actual world w_0 but it is true at the non-actual world w_1. And the truth-value of a sentence also depends on the reference function. So a sentence is true or false *at a world* given a reference function.

A sentence of the form <NAME is NOUN> has these truth-conditions:

the value of <NAME is NOUN> at world w given f
$= f(\text{NAME}, w) \in f(\text{NOUN}, w)$.

For example,

the value of "Betty is a woman" at w_1 given f
$= f(\text{"Betty"}, w_1) \in f(\text{"woman"}, w_1)$
$= B \in \{B, D\}$
= true.

A sentence of the form <NAME is ADJ> has these truth-conditions:

the value of <NAME is ADJ> at w given f
$= f(\text{NAME}, w) \in f(\text{ADJ}, w)$.

For example,

the value of "Betty is happy" at w_1 given f
$= f(\text{"Betty"}, w_1) \in f(\text{"happy"}, w_1)$
$= B \in \{A, B, D\}$
= true;

the value of "Betty is happy" at w_4 given f
$= f(\text{"Betty"}, w_4) \in f(\text{"happy"}, w_4)$
$= B \in \{A, D\}$
= false.

A sentence of the form <NAME_1 VERB NAME_2> has these truth-conditions:

the value of <NAME_1 VERB NAME_2> at w given f
$= (\, f(\text{NAME}_1, w), f(\text{NAME}_2, w)\,) \in f(\text{VERB}, w)$.

For example,

the value of "Allan loves Betty" at w_1 given f
$= (\, f(\text{"Allan"}, w_1), f(\text{"Betty"}, w_1)\,) \in f(\text{"love"}, w_1)$
$= (A, B) \in \{(A, B), (B, A), (C, D)\}$
= true;

the value of "Allan loves Betty" at w_4 given f
= (f("Allan", w_4), f("Betty", w_4)) $\in$ f("love", w_4)
= (A, B) $\in$ {(A, D), (D, A), (C, B)}
= false.

2.4 Modalities

One of the great advantages of possible worlds semantics is that it treats the classical modes of necessity and possibility in terms of *quantification* over worlds. This means that we analyze necessity and possibility in terms of the *existential quantifier* (there exists x such that . . . x . . .) and the *universal quantifier* (for every x, it is the case that . . . x . . .). When we say we are quantifying over worlds, we interpret the existential quantifier as talking about worlds (there exists a world x such that . . . x . . .) and likewise we interpret the universal quantifier as talking about worlds (for every world x, it is the case that . . . x . . .). For precision, we need to distinguish between two ways to use the modes of necessity and possibility: *de dicto* and *de re*. This distinction is subtle, and sometimes hard to see. We won't go into a deep discussion of the metaphysics. We only sketch some truth-conditions.

De Dicto Necessity. A de dicto necessity is a statement about the modality of a *sentence*. A de dicto necessity has the form <It is necessary that SENTENCE>. For instance, "It is necessary that Charlie is sad" is de dicto. It says that the *sentence* "Charlie is sad" has a property, namely, the property of being necessarily true. A sentence is necessarily true if, and only if, it is true at every world. Thus

the value of <It is necessary that SENTENCE> at w given f
= (for every world v)(SENTENCE is true at v given f).

For example, in our simple model, the sentence "Charlie is sad" is true at every world:

the value of "It is necessary that Charlie is sad" at w given f
= (for every world v)("Charlie is sad" is true at v given f)
= "Charlie is sad" is true at w_1, w_2, w_3, w_4 given f
= true.

To see the difference between truth and necessary truth, consider "Betty is happy". It is true at w_1. But it is not *necessarily* true at w_1 (or any other world). The sentence "It is necessary that Betty is happy" is false at every world; she is not happy at w_4.

De Dicto Possibility. A de dicto possibility has the form <It is possible that SENTENCE>. For instance, "It is possible that Betty is a man" is de dicto. It says that the *sentence* "Betty is a man" has a property, namely, the property of being possibly true. A sentence is possibly true if, and only if, it is true at some world. Thus

the value of <It is possible that SENTENCE> at *w* given *f*
= (for some world *v*)(SENTENCE is true at *v* given *f*).

For example,

the value of "It is possible that Betty is a man" at *w* given *f*
= (for some world *v*)("Betty is a man" is true at *v* given *f*)
= "Betty is a man" is true at either w_1, or w_2, or w_3, or w_4 given *f*
= false.

To see the difference between truth and possible truth, consider "Betty is sad". It is not true at w_1. But it is *possibly* true at w_1 (and at every other world too). The sentence "It is possible that Betty is sad" is true at every world, since she is sad at w_4.

De Re Necessity. A de re necessity is a statement about a thing. Consider <NAME is necessarily ADJ>. Any instance of that syntactic form says that a thing has some property essentially. It does not say that a sentence necessarily has truth (or falsity), but rather that a thing necessarily has a property. It says that at every world at which the thing exists, it has the property. Specifically,

the value of <NAME is necessarily ADJ> at *w* given *f*
= for every world *v* at which NAME exists,
<NAME is ADJ> is true at *v* given *f*.

the value of <NAME is necessarily ADJ> at *w* given *f*
= (for every world *v*)
(if *f*(NAME, *v*) ∈ *f*("thing", *v*), then *f*(NAME, *v*) ∈ *f*(ADJ, *v*)).

As an illustration of the difference between de dicto necessity and de re necessity, compare "It is necessary that Eric is a man" with "Eric is necessarily a man". The de dicto "It is necessary that Eric is a man" says that "Eric is a man" is true at every world. But it is not, since Eric does not exist at worlds w_3 and w_4. Hence Eric is not a man (or anything else) at those worlds. But the de re "Eric is necessarily a man" says that at every world at which Eric exists, Eric is a man. And this is true; for at the worlds w_1 and w_2, Eric is a man.

De Re Possibility. A de re possibility is a statement about a thing. Consider <NAME is possibly ADJ>. Any instance of that syntactic form says that a thing has some property in a possible way. It does not say that a sentence possibly has truth (or falsity), but rather that a thing possibly has a property. It says that there is some world at which the thing exists and has that property. Specifically,

the value of <NAME is possibly ADJ> at *w* given *f*
= there is some world *v* at which NAME exists,
<NAME is ADJ> is true at *v* given *f*.

the value of <NAME is possibly ADJ> at *w* given *f*
= (there is some world *v*)
$((f(\text{NAME}, v) \in f(\text{"thing"}, v))$ & $(f(\text{NAME}, v) \in f(\text{ADJ}, v)))$.

2.5 Intensions

The *intension* of a word is a function that associates every world with the referent of that word at that world. We let IN be the intension function. Thus

$$\text{IN}(\text{"Allan"}) = \{(w_1, A), (w_2, A), (w_3, A), (w_4, A)\};$$

$$\text{IN}(\text{"Diane"}) = \{(w_1, D), (w_2, D), (w_3, D), (w_4, D)\}.$$

Our models determine the intensions of some nouns, adjectives, and verbs as follows:

$$\text{IN}(\text{"man"}) = \{ (w_1, \text{man at } w_1), (w_2, \text{man at } w_2), (w_3, \text{man at } w_3), (w_4, \text{man at } w_4)\};$$

$$= \{ (w_1, \{A, C\}), (w_2, \{A, C, E\}), (w_3, \{A, C\}), (w_4, \{A, C\})\}.$$

Notice that an intension is a function whose output is a function. An intension has to be supplied with inputs *twice* before yielding a final output. For example,

$$\text{IN}(\text{"Allan"}) = \{(w_1, A), (w_2, A), (w_3, A), (w_4, A)\};$$

$$(\text{IN}(\text{"Allan"}))(w_1) = A.$$

The truth-conditions of sentences can easily be given in terms of intensions rather than in terms of a 2-place reference function. Using intensions, a sentence of form <NAME is NOUN> has these truth-conditions:

<NAME is NOUN> is true at world *w* given *f*
$= ((f(\text{NAME}))(w) \in (f(\text{NOUN}))(w))$.

2.6 Propositions

At every world, a given sentence has a truth-value. We can thus define a function that maps each (sentence, world) pair onto the truth-value (0 for false and 1 for true) of the sentence at that world. Alternatively, and more simply, we can let the intension of a sentence be a function that associates every world with the truth-value of the sentence at that world. We have a set of worlds W. The set of characteristic functions over W is $\{ f \mid f: W \rightarrow \{0, 1\}\}$. Each of these characteristic functions is an intension.

Propositions. Some philosophers identify the meaning of a sentence with its intension. And the meaning of a sentence is usually said to be the *proposition* that is expressed by the sentence. The intension of a sentence is thus the proposition expressed by the sentence. Accordingly, a *proposition* is a function that associates each world with a truth-value. If a proposition associates a world *w* with 1, then it is true at *w*; if 0, it is false at *w*. Let's use square brackets around a sentence to denote the proposition expressed by that sentence. Here are some examples:

[Charlie is sad] $= \{(w_1, 1), (w_2, 1), (w_3, 1), (w_4, 1)\}$;

[Charlie is happy] $= \{(w_1, 0), (w_2, 0), (w_3, 0), (w_4, 0)\}$;

[Allan loves Betty] $= \{(w_1, 1), (w_2, 1), (w_3, 1), (w_4, 0)\}$.

A proposition that is true at every world is necessarily true; it is a necessary truth. A proposition that is false at every world is a necessary falsehood. For example, "Charlie is sad" is a necessary truth while "Charlie is happy" is a necessary falsehood. Charlie isn't happy at any world. Poor Charlie.

Finally, some philosophers prefer to identify the proposition expressed by a sentence with the set of worlds at which a sentence is true. Thus

[Charlie is sad] $= \{w_1, w_2, w_3, w_4\}$;

[Charlie is happy] $= \{\}$;

[Allan loves Betty] $= \{w_1, w_2, w_3\}$.

The conception of propositions as sets of possible worlds is useful in discussions of probability. One way to define the probability of a sentence (relative to some non-empty set of worlds) is to identify it with the number of worlds at which the sentence is true divided by the number of worlds. For example, in our model, there are 4 possible worlds; in three of them, Allan loves Betty; hence the probability that Allan loves Betty is 3/4.

3. Modal Semantics with Counterparts

3.1 The Counterpart Relation

An alternative to our first version of possible worlds semantics is David Lewis's counterpart theory (1968). Counterpart theory says that worlds don't share individuals. Worlds don't overlap. Each individual is in exactly one world. As you might expect, counterpart theory solves some problems while raising others. But we're not here to judge. We don't need to go into the metaphysical issues here. Our purpose here is merely to develop some of the mathematical machinery behind the metaphysics.

According to counterpart theory, reality is the modal structure (V, W, I, δ, C, f). As before, V is an ordered tuple of vocabulary item (words). W is a set of possible worlds. I is a set of individuals. The items δ and C are specific to counterpart theory. Item δ is a function that associates every world with the set of individuals in that world. The item C is the counterpart relation. Finally, f is the reference function. We'll discuss these last three items in detail.

We consult δ to determine whether an individual is in a world. That is, x is in world w iff x is in $\delta(w)$. The item δ can be used to define a worldmate relation on individuals. We say *x is a worldmate of y* iff there is some world w such that x is in $\delta(w)$ and y is in $\delta(w)$. According to counterpart theory, worlds are non-empty; they do not overlap (they share no individuals); and they exhaust the set of possible individuals. Hence the worldmate relation is an equivalence relation that partitions the set of individuals into equivalence classes. Each equivalence class belongs to a world. It is all the things in that world and no other. The function δ maps each world onto its equivalence class of worldmates.

The counterpart relation associates an individual with its counterparts. You are *represented* at other worlds by your counterparts. The counterpart relation is a relation of similarity. Roughly, your counterpart at some world is the thing in that world that is maximally similar to you. On this view, your counterpart in your world is you. You are maximally similar to yourself. Since each thing is a counterpart of itself, the counterpart relation is reflexive. But what about symmetry? It might seem obvious that if x is a counterpart of y, then y is a counterpart of x. But we need not require that. And we need not require transitivity. Further, an individual at one world can have many counterparts at another world.

Finally, we come to the reference function f. This function associates words with their referents. According to counterpart theory, a proper name refers to a single thing that exists in exactly one world. We use "Al Gore" to refer to Al Gore, who lives in our world. We use "Sherlock Holmes" to refer to Sherlock Holmes, who lives in another world. A predicate refers to a single extension that spans worlds. For instance, the extension of "detective" includes all actual detectives as well as non-actual possible detectives, like Sherlock Holmes. The extension of "loves" includes all pairs of lovers at all possible worlds. It includes (Romeo, Juliet) as well as pairs of actual lovers. Note that extensions are not split up across worlds. One property is the same at all worlds.

3.2 A Sample Model for Counterpart Theoretic Semantics

For the sake of illustration, we discuss a single modal structure (V, W, I, C, δ, *f*). As before, the vocabulary V is the following ordered tuple of words:

("Allan", "Betty", "Charlie", "Diane", "Eric",
"human", "man", "woman", "loves", "happy", "sad").

The set of worlds W is { w_1, w_2, w_3, w_4}.

Among these worlds, we take w_1 to be the actual world.

The set of individuals I is

{ A_1, B_1, C_1, D_1, A_2, B_2, C_2, D_2, E_2, A_3, B_3, C_3, A_4, B_4, C_4, D_4}.

There is an inclusion function δ. The inclusion function associates each world with the set of individuals in that world. We define δ like this:

$\delta(w_1) = \{A_1, B_1, C_1, D_1\}$;
$\delta(w_2) = \{A_2, B_2, C_2, D_2, E_2\}$;
$\delta(w_3) = \{A_3, B_3, C_3\}$;
$\delta(w_4) = \{A_4, B_4, C_4, D_4\}$.

Counterparts are individuals with the same letter. For example, the members of {A_1, A_2, A_3, A_4} are all counterparts of one another. Likewise all the B's are counterparts of one another; all the C's are; all the D's are; and all the E's are too.

There is a reference function *f* given in Table 4.3. There are two things to notice about this reference function. The first is that it looks a lot like the extensional reference function in Table 4.1. Counterpart theory is more extensional. The second thing to notice is that most of our names refer to things in w_1. This is as expected, since w_1 is the actual world in our example. We usually refer to actual things and less commonly to non-actual things. Our example has only one name, "Eric", that refers to a non-actual thing.

Term	Referent
"Allan"	A_1
"Betty"	B_1
"Charlie"	C_1
"Diane"	D_1
"Eric"	E_2
"human"	$\{A_1, B_1, C_1, D_1, A_2, B_2, C_2, D_2, E_2, A_3, B_3, C_3, A_4, B_4, C_4, D_4\}$
"man"	$\{A_1, C_1, A_2, C_2, E_2, A_3, C_3, A_4, C_4\}$
"woman"	$\{B_1, D_1, B_2, D_2, B_3, B_4, D_4\}$
"loves"	$\{(A_1, B_1), (B_1, A_1), (C_1, D_1), (A_2, B_2), (B_2, A_2), (C_2, D_2), (D_2, E_2), (E_2, D_2), (A_3, B_3), (B_3, A_3), (C_3, B_3), (A_4, D_4), (D_4, A_4), (C_4, B_4)\}$
"happy"	$\{A_1, B_1, D_1, A_2, B_2, D_2, E_2, A_3, B_3, A_4, D_4\}$
"sad"	$\{C_1, C_2, C_3, C_4, B_4\}$

Table 4.3 Our Lewisian reference function.

3.3 Truth-Conditions for Non-Modal Statements

As before, sentences are true or false at worlds. The truth-value of a sentence depends on the world and the reference function. We thus say a sentence is true or false at a world *w* given a reference function *f*.

The idea behind counterpart theory is that when we are talking about what is going on at some world, we are restricting our attention to that world. (Technically speaking, we are restricting our quantifiers to that world.) For example, to say that "Al Gore lost" at our world is to say that, if we just look at the things in our world, one of them is Al Gore, and he lost. Designating our world with the symbol "@", that is to say that there exists some thing x in $\delta(@)$ such that x is Al Gore and x lost. More generally:

the value of <NAME is ADJ> at world w
= (there is some x in $\delta(w)$)(x = NAME & x is ADJ).

For full precision, we need to use the reference function:

the value of <NAME is ADJ> at world w given f
= (there exists $x \in \delta(w)$)($x = f$(NAME) & $x \in f$(ADJ)).

For example,

the value of "Betty is happy" at w_1 given f
$= (\text{there exists } x \in \delta(w_1))(x = f(\text{"Betty"}) \ \& \ x \in f(\text{"happy"}))$
$= (B_1 \in \delta(w_1))(B_1 = B_1 \ \& \ B_1 \in \{A_1, B_1, \ldots D_4\})$
= true.

Analogously,

the value of <NAME is NOUN> at world w given f
$= (\text{there exists } x \in \delta(w))(x = f(\text{NAME}) \ \& \ x \in f(\text{NOUN})).$

We illustrate this with "Betty is a woman", which is true at w_1 given f:

the value of "Betty is a woman" at w_1 given f
$= (\text{there exists } x \in \delta(w_1))(x = f(\text{"Betty"}) \ \& \ x \in f(\text{"woman"}))$
$= (B_1 \in \delta(w_1))(B_1 = B_1 \ \& \ B_1 \in \{B_1, D_1, \ldots D_4\})$
= true.

But observe that "Betty is a woman" is false at w_4 given f. The name "Betty" refers to the actual woman B_1, and B_1 is not in $\delta(w_4)$. Hence B_1 cannot be a woman in that other world. B_1 is a woman only in world w_1. Now consider Eric. We can talk about Eric's properties at w_2. That's the world in which he exists. Thus

the value of "Eric is a man" at w_2 given f
$= (\text{there exists } x \in \delta(w_2))(x = f(\text{"Eric"}) \ \& \ x \in f(\text{"man"}))$
$= (E_2 \in \delta(w_2))(E_2 = E_2 \ \& \ E_2 \in \{A_1, \ldots E_2, \ \ldots C_4\})$
= true.

For relational statements the truth-conditions are

the value of <NAME₁ VERB NAME₂> at world w given f
$= (\text{there exists } x \in \delta(w))(x = f(\text{NAME}_1) \ \&$
$(\text{there exists } y \in \delta(w))(y = f(\text{NAME}_2) \ \&$
$(x, y) \in f(\text{VERB}))).$

For example,

the value of "Allan loves Betty" at world w_1 given f
$= (\text{there exists } A_1 \in \delta(w_1))(A_1 = f(\text{"Allan"}) \ \&$
$(\text{there exists } B_1 \in \delta(w_1))(B_1 = f(\text{"Betty"}) \ \&$
$(A_1, B_1) \in \{(A_1, B_1), \ldots (C_4, B_4)\})$
= true.

3.4 Truth-Conditions for Modal Statements

Consider the de dicto statement "It is possible that Al Gore wins". Since Al Gore might have won, you might think that it is possible that Al Gore wins. But it is not possible that *Al Gore* wins. Why not? Because to say that it is possible that Al Gore wins is to say that there is some world in which Al Gore wins. But, according to counterpart theory, Al Gore is in exactly one world, and in that world, Al Gore loses. More formally,

> the value of "It is possible that Al Gore wins" at our world @ iff
> (there is some world v)
> ("Al Gore wins" is true at v given f).

But that is false. So bear in mind: counterpart theory says no individual exists in more than one world. This affects the truth-values of de dicto statements.

De Dicto Necessity. For de dicto necessity, we have

> the value of <It is necessary that NAME is ADJ> at w given f
> = (for every world v)(<NAME is ADJ> is true at v given f).

As a more complex example, consider

> the value of <It is necessary that all NOUN are ADJ> at w given f
> = (for every world v)(<All NOUN are ADJ> is true at v given f).

De Dicto Possibility. For de dicto possibility, we have

> the value of <It is possible that NAME is ADJ> at w given f
> = (for some world v)(<NAME is ADJ> is true at v given f).

As a more complex example, consider

> the value of <It is possible that all NOUN are ADJ> at w given f
> = (for some world v)(<All NOUN are ADJ> is true at v given f).

Consider the de re statement "Al Gore might have won". This is true iff there is some world at which there is a counterpart of our Al Gore and the counterpart wins. Why the counterpart? Because our Al Gore did not win. He does not have the property of winning. Nor does he have the property of winning-at-some-other-world. Winning is winning – it is the same property from world to world; it has a single extension that spans worlds. Thus

> the value of "Al Gore might have won" at our world @
> = (there is some other world v)
> (there is some x in v)(x is a counterpart of Al Gore & x wins).

The counterpart relation is used extensively in de re statements. We can use $x \approx y$ to symbolize x is a counterpart of y (that is, $C(x, y)$ in our model).

De Re Necessity. We can now define de re necessity. This modality involves the use of the counterpart relation. For de re necessity, we have

the value of <NAME is necessarily ADJ> at w given f
= for every counterpart of NAME, that counterpart is ADJ.

Putting this into symbols we get

the value of <NAME is necessarily ADJ> at w given f
= (for every world v)
(for every $x \in \delta(v)$)(if $x \approx f(\text{NAME})$ then $x \in f(\text{ADJ})$).

As a more complex example, consider

the value of <All NOUN are necessarily ADJ> at w given f
= (for every x in w)
(if x is a NOUN, then
(for *every* counterpart of x, that counterpart is ADJ));

the value of <All NOUN are necessarily ADJ> at w given f
= (for every x in w)
(if x is a NOUN, then
(for every world v)
(for every y in v)(if y is a counterpart of x, then y is ADJ));

the value of <All NOUN are necessarily ADJ> at w given f
= (for every $x \in \delta(w)$)
(if $x \in f(\text{NOUN})$ then
(for every world v)
(for every $y \in \delta(v)$)(if $y \approx x$ then $y \in f(\text{ADJ})$)).

De Re Possibility. For de re possibility, we have

the value of <NAME is possibly ADJ> at w given f
= for some counterpart of NAME, that counterpart is ADJ;

the value of <NAME is possibly ADJ> at w given f
= (for some world v)
(for some $x \in \delta(v)$)($x \approx f(\text{NAME})$ & $x \in f(\text{ADJ})$).

As a more complex example, consider

the value of <All NOUN are possibly ADJ> at w given f
  = (for every x in w)
    (if x is a NOUN, then
      (for *some* counterpart of x, that counterpart is ADJ));

the value of <All NOUN are possibly ADJ> at w given f
  = (for every x in w)
    (if x is a NOUN, then
      (for some world v)
        (for some y in v)(y is a counterpart of x & y is ADJ));

the value of <All NOUN are possibly ADJ> at w given f
  = (for every $x \in \delta(w)$)
    (if $x \in f(\text{NOUN})$ then
      (for some world v)
        (for some $y \in \delta(v)$)($y \approx x$ & $y \in f(\text{ADJ})$)).

5

PROBABILITY

1. Sample Spaces

Experiment. An experiment is any change that actualizes one of many possible outcomes. A first example of an experiment is throwing a cubical die. An outcome is the number on the top side of the die. There are six possible outcomes, and throwing the die ensures that exactly one of them is actualized. A second example of an experiment is picking a card out of a deck of cards. Since there are 52 cards in a deck, there are 52 possible outcomes; selecting a single card means that exactly one outcome is actualized. A third example of an experiment is a lottery. If 1,000,000 tickets are issued, then there are 1,000,000 possible outcomes. Selecting the winning number means that exactly one outcome is actualized. A fourth example is throwing basketballs from the foul line. There are two possible outcomes: the ball goes through the hoop, or it does not. Throwing the ball at the basket actualizes exactly one of these possibilities.

Sample Space. The sample space of an experiment is the set of possible outcomes. When throwing a die, the sample space is the set of six sides of the die. These sides make the set $\{d_1, d_2, d_3, d_4, d_5, d_6\}$. But we can also just represent each side by its number, so the sample space for a cubical die can also be represented just by the set of numbers {1, 2, 3, 4, 5, 6}. When picking a card, the sample space is the set of pairs (value, suit) where value is the set {Ace, 2, . . . 10, Jack, Queen, King} and suit is the set {Club, Heart, Spade, Diamond}. When shooting baskets, the sample space is {hit, miss}.

Event. Although the term *event* is usually used in philosophy to mean a single particular occurrence, in probability theory, an event is a collection of possibilities. An event, relative to an experiment, is a subset of the sample space of the experiment. It is a subset of the set of possible outcomes of the experiment. Consider selecting a card from a deck:

Event_1 = { (Ace, Club) };

Event_2 = { (2, Heart), (3, Club), (King, Heart) };

Event_3 = { (Ace, Club), (King, Club), (Queen, Club), (Jack, Club), (10, Club) };

Event_4 = $\{ x \mid x$ is a heart $\}$;

Event_5 = $\{ x \mid x$ is a face card $\}$;

$\text{Event}_6 = \{ x \mid x \text{ is not a face card} \}$.

Since the entire sample space is a subset of the sample space, the entire sample space is an event. Likewise, since the empty set is a subset of the sample space, the empty set is an event (sometimes called the *null event* or the *empty event*). Thus

$\text{Event}_7 = \{ x \mid x \text{ is a card} \}$;

$\text{Event}_8 = \{\}$.

2. Simple Probability

Probability of an Event. Suppose an experiment has a *finite* sample space S, and that the event E is a subset of S. Suppose further that every outcome of S is *equally likely*. For example, if a coin is fair, then heads is just as likely as tails. However, if a die is loaded, then not every outcome is equally likely; some are favored over others.

At this point, we're concerned only with experiments with finitely many, equally likely outcomes. For such experiments, the probability of an event is the number of outcomes in the event divided by the number of possible outcomes of the experiment. More technically, it is the cardinality of the event divided by the cardinality of the sample space. Recall that we write the cardinality of a set S as $|S|$. We write the probability of an event E as P(E). In symbols, this probability is

$P(E) = |E| / |S|$.

For example, suppose a bag contains 10 marbles. Three are white, and seven are black. What is the probability of pulling out a white marble? The sample space is

$S = \{w_1, w_2, w_3, b_4, b_5, b_6, b_7, b_8, b_9, b_{10} \}$.

The event is

$W = \{w_1, w_2, w_3\}$.

Hence the probability is $|W| / |S| = 3 / 10$. This corresponds to the idea that 3 out of 10 choices are white marbles; the other 7 choices are black marbles.

As another example, suppose the experiment is three tosses of a fair coin. If we let "h" represent heads and "t" represent tails, then the sample space is

S = {hhh, hht, hth, htt, thh, tht, tth, ttt}.

What is the probability of getting exactly two heads? The event is

E = {hht, hth, thh}.

Hence the probability of getting exactly two heads is |E| / |S| = 3 / 8.

As a third example, suppose the sample space is a deck of cards and the experiment is to select a single card from the deck. What is the probability of selecting a heart? Here

S = { x | x is a card };

E = { x | x is a heart }.

Since there are 52 cards and 13 hearts, the probability is 13/52 = 1/4.

There are two trivial probabilities. First, since the sample space is an event (that is, it is a subset of itself), the probability of the sample space is

P(S) = |S| / |S| = 1.

For example, the probability of drawing a card from a deck of cards is 1. The second trivial probability is that of the null event. The null event is the empty set {}. Its cardinality is 0. So the probability of the null event is

P({}) = 0 / |S| = 0.

For example, if the experiment is drawing a card from a deck of cards, then the null event is not drawing a card. The probability of not drawing a card is 0.

The sample space of an experiment is a set. It therefore has a power set. If S is the sample space of an experiment, then the power set of S is the set of all events in the experiment. If pow S is the power set of S, then S is the maximal set in that power set and {} is the minimal set. As we mentioned, the maximal set in pow S has the maximal probability 1 and the minimal set in pow S has the minimal probability 0. But the sets (the events) in between {} and S have probabilities between 0 and 1. Recall that the sign ⊂ denotes a *proper* subset – a subset that is not equal to the set. Put symbolically,

P({}) = 0;

if {} ⊂ E ⊂ S then 0 < P(E) < 1;

P(S) = 1.

3. Combined Probabilities

We are often interested in combining the probabilities of different events. Since events are sets, they have intersections, unions, and complements. So we can ask about the probabilities of intersections, unions, and complements of events.

Intersections of Events. The intersection of two events is usually expressed by a conjunction: what is the probability of an outcome being *both* in E_1 *and* in E_2? For example, what is the probability of drawing a card that is a heart and a face card? We can express this as the intersection of two events:

$$E_1 = \{ x \mid x \text{ is a heart } \};$$

$$E_2 = \{ x \mid x \text{ is a face card}\}.$$

The probability of drawing a card that is a heart and a face card is the probability of

$$E_3 = \{ x \mid x \text{ is in } E_1 \text{ and } x \text{ is in } E_2 \} = E_1 \cap E_2.$$

We are thus interested in $P(E_1 \cap E_2)$. This is defined as expected:

$$P(E_1 \cap E_2) = |E_1 \cap E_2| \, / \, |S|.$$

For our example, $E_1 \cap E_2 = \{$ (Jack, Heart), (Queen, Heart), (King, Heart)$\}$. So the probability of drawing a face card of hearts is 3/52.

Unions of Events. The union of two events is usually expressed by a disjunction: what is the probability of the outcome being *either* in E_1 *or* in E_2? For example, what is the probability of a card being either a heart or a face card? This is as expected:

$$P(E_1 \cup E_2) = |E_1 \cup E_2| \, / \, |S|.$$

For unions, the computation of the cardinality of $|E_1 \cup E_2|$ is trickier than it might appear. To compute $|E_1 \cup E_2|$, we can't just add $|E_1|$ to $|E_2|$. Why not? Because some outcomes might be in both events. And they would be counted twice if we just added $|E_1|$ to $|E_2|$, which would be incorrect. Consider the cards. Some cards are both hearts and face cards. When we count the number of cards that are either hearts or face cards, we don't want to count them twice. Hence, when computing the number of outcomes in the union of E_1 and E_2, we need to add the outcomes in E_1 to the outcomes in E_2, and to subtract the outcomes in both E_1 and E_2. Our formula for $|E_1 \cup E_2|$ is

$$|E_1 \cup E_2| = |E_1| + |E_2| - |E_1 \cap E_2|.$$

Hence

$$P(E_1 \cup E_2) = (|E_1| + |E_2| - |E_1 \cap E_2|) / |S|.$$

Now, we can divide each term in the numerator by |S| to get a nicer formula expressing the probability of the union only in terms of other probabilities:

$$P(E_1 \cup E_2) = |E_1| / |S| + |E_2| / |S| - |E_1 \cap E_2| / |S|;$$

$$P(E_1 \cup E_2) = P(E_1) + P(E_2) - P(E_1 \cap E_2).$$

For example, the probability of picking either a heart or a face card is

$$P(\text{Heart} \cup \text{Face}) = P(\text{Heart}) + P(\text{Face}) - P(\text{Heart} \cap \text{Face});$$

$$P(\text{Heart} \cup \text{Face}) = 13/52 + 12/52 - 3/52 = 22/52.$$

When two events E_1 and E_2 are *mutually exclusive*, then they have an empty intersection. The probability of an outcome being in both E_1 and E_2 is zero. Hence, if E_1 and E_2 are mutually exclusive,

$$P(E_1 \cup E_2) = P(E_1) + P(E_2).$$

Complements of Events. The complement of an event E, relative to a sample space S, is the set of outcomes that are in S, but not in E. It is

$$E^* = S - E = \{ x \mid x \text{ is in S \& } x \text{ is not in E} \}.$$

The probability is straightforward:

$$P(E^*) = (|S| - |E|) / |S| = (|S| / |S|) - (|E| / |S|) = 1 - P(E).$$

For example, what is the probability of not drawing a face card from a deck? We refer to the event of not drawing a face card as *not Face*. Since there are 52 cards in a deck, of which exactly 12 are face cards, we have

$$P(\text{not Face}) = (|\text{Cards}| - |\text{Faces}|) / |\text{Cards}| = (52 - 12) / 52 = 40 / 52.$$

Alternatively, since there are 12 face cards in a deck, P(Face) = 12/52. Thus

$$P(\text{not Face}) = 1 - P(\text{Face}) = 1 - 12/52 = 40 / 52.$$

4. Probability Distributions

So far we've been talking about experiments whose sample spaces are finite and in which all outcomes are equally likely. But it often happens that the outcomes of an experiment are not all equally likely. For example, consider shooting baskets from the foul line. The experiment has two possible outcomes: hit or miss. Presumably, these are not equally likely; the outcome depends on the skill of the player. If the player is good, we expect that she'll hit many more foul shots than she'll miss. Or, consider a loaded die. A die might be rigged so that even numbers are twice as likely to occur as odd numbers.

Probability Distribution. A probability distribution (aka a probability *assignment*) is a function P from a sample space S to the set of real numbers between 0 and 1 such that for any outcome x in S, P(x) is the probability of x. A *real number* is defined by a sequence of digits, followed by a decimal point, followed by a finite or infinite sequence of digits. For example, 0.5 is a real number; 0.3333. . . is a real number; 1.0 is a real number.

There are two necessary constraints on probability distributions. First of all, every outcome in the sample space has to have some probability. Its probability is 0 if it cannot occur, and 1 if it must occur. Hence for every outcome x in S, the probability of x is between 0 and 1. In symbols,

$$\text{for every } x \in \text{S}, 0 \leq \text{P}(x) \leq 1.$$

Secondly, an experiment by definition has an outcome. So it is necessary that some outcome occurs. The probability that some outcome occurs is 1. But this is just the sum of the probabilities of the individual outcomes. Thus the sum, for every outcome x in the sample space S, of the probability of x, is 1. Recalling the notation for sums introduced in Chapter 1, section 16, we write it like this:

$$\sum_{x \in \text{S}} \text{P}(x) = 1.$$

When each outcome of an experiment is equally likely, we have a special case. Suppose there are n outcomes. These are the outcomes o_1 through o_n. Thus the sample space S = {o_1, . . . o_n}. We know that the sum, for all n, of P(o_n) is 1. That is,

$$\text{P}(o_1) + \ldots . + \text{P}(o_n) = 1.$$

Since all these probabilities are equal, we can substitute P(o_1) for every other P(o_i). We thus have that $n \cdot \text{P}(o_1) = 1$ and $\text{P}(o_1) = 1/n$. Since every other outcome has the same probability as o_1, it follows that for every outcome o_i, $\text{P}(o_i) = 1/n$.

For example, consider a fair cubical die. The outcomes are d_1 to d_6. Each is equally likely and the sum of them all is 1. We thus have:

$$P(d_1) = P(d_2) = P(d_3) = P(d_4) = P(d_5) = P(d_6);$$

$$P(d_1) + P(d_2) + P(d_3) + P(d_4) + P(d_5) + P(d_6) = 1.$$

And by substituting $P(d_1)$ for every other probability we get:

$$P(d_1) + P(d_1) + P(d_1) + P(d_1) + P(d_1) + P(d_1) = 1,$$

so that $6 \cdot P(d_1) = 1$, and hence $P(d_1) = 1/6$. Since the probabilities of all the outcomes are equal, it follows that the probability of each outcome is 1/6.

Given a probability distribution for some sample space S, we want to know the probability of an event. We've defined probabilities for *members* of S, but an event is a *subset* of S. How do we define the probability of the event? We start by extending P from members of S to *unit events* over S. A unit event is a unit set. If x is an outcome in S, then $\{x\}$ is a unit event. The extension is easy: $P(\{x\}) = P(x)$. To continue our extension of P to events, we reason like this: an event E is a set of outcomes $\{o_1, \ldots o_n\}$. We can express E as the union of unit events, one for each outcome. That is,

$$E = \{o_1, \ldots o_n\} = \{o_1\} \cup \ldots \cup \{o_n\}; \text{ and thus}$$

$$P(E) = P(\{o_1, \ldots o_n\}) = P(\{o_1\} \cup \ldots \cup \{o_n\}).$$

These unit events don't overlap at all. Their intersections are always empty, so the probability of their intersections is always 0. Therefore we don't need to worry about counting any outcome twice when computing $P(\{o_1\} \cup \ldots \cup \{o_n\})$. We just add:

$$P(E) = \text{the sum, for } i \text{ varying from 1 to } n, \text{ of } P(o_i).$$

More generally, P(E) = the sum, for all x in E, of $P(x)$. Put in notation, we have

$$P(E) = \sum_{x \in E} P(x) \,.$$

Calculating probability distributions can be tricky. Suppose a die is loaded so that the even numbers are twice as likely as the odd numbers. Besides that, there is no bias. Any even number is as likely as any other even number and any odd number is as likely as any other odd number. We have the following:

$$P(\text{Even}) = 2 \cdot P(\text{Odd}).$$

We know that the die has to come up either even or odd, so that

$$P(\text{Even}) + P(\text{Odd}) = 1.$$

By substituting, we learn that

$$2 \cdot P(\text{Odd}) + P(\text{Odd}) = 3 \cdot P(\text{Odd}) = 1.$$

Thus P(Odd) = 1/3 and P(Even) = 2 · P(Odd) = 2/3. Of course, we aren't done. We need the probabilities for the individual outcomes. We know that

$$P(\text{Odd}) = P(d_1) + P(d_3) + P(d_5) = 1/3.$$

Since any odd number is as likely as any other, we know that

$$P(d_1) = P(d_3) = P(d_5).$$

Let $P(d_1)$ be x. We have $x + x + x = 1/3$. Thus $3x = 1/3$ and $x = 1/9$. Since the probabilities of all odd numbers are equal, we have $P(d_1) = P(d_3) = P(d_5) = 1/9$.

We can perform analogous calculations for the even numbers:

$$P(\text{Even}) = P(d_2) + P(d_4) + P(d_6) = 2/3.$$

Letting $P(d_2) = x$ we have $3x = 2/3$ and thus $P(d_2) = P(d_4) = P(d_6) = 2/9$.

Now let's consider another case involving the biased die. For this die, even numbers are twice as likely as odd numbers. What is the probability of getting a number greater than 3? The set of outcomes in this event is $\{d_4, d_5, d_6\}$. Thus we have the probability

$$P(\{d_4, d_5, d_6\}) = P(d_4) + P(d_5) + P(d_6) = 2/9 + 1/9 + 2/9 = 5/9.$$

You might try to figure out the probability of getting a number less than 5.

5. Conditional Probabilities

5.1 Restricting the Sample Space

We often want to know the probability of one event given that another event has already happened. For example, suppose you're rolling a fair six-sided die. The probability of rolling an even number is P(Even) = 3/6. The probability of rolling a number less than 4 is P(Less Than 4) = 3/6. What is the probability of rolling an even number given that you've rolled a number less than 4? For the sake of convenience, instead of talking about rolling an outcome in $\{d_1, d_2, d_3\}$, we'll just talk about the probability of rolling a number in {1, 2, 3}. The number n indicates the outcome d_n.

What is the probability of rolling an even number given that you've rolled a number in {1, 2, 3}? You can see just by counting that there is 1 way to roll an even number in {1, 2, 3}, and that there are 3 possible ways to roll a number in {1, 2, 3}. Since the die is fair, this means that your chance of rolling an even number in {1, 2, 3} is 1 in 3. Your chance is the number of ways to roll an even number in {1, 2, 3} divided by the number of ways to roll any number in {1, 2, 3}. That is, the probability of rolling an even number given that you've rolled in {1, 2, 3} is

$$\text{P(Roll is even given roll is in } \{1, 2, 3\})$$
$$= \frac{\text{ways to roll an even number in } \{1, 2, 3\}}{\text{ways to roll a number in } \{1, 2, 3\}}.$$

You roll an even number in {1, 2, 3} iff you roll an even number, and you roll a number in {1, 2, 3}. You roll an even number in {1, 2, 3} iff you roll a number in the intersection of the set of even numbers with the set {1, 2, 3}. The intersection indicates that you've restricted the sample space to {1, 2, 3}. The number of ways to roll an even number within this restricted sample space is the cardinality of (Even $\cap$ {1, 2, 3}). And the number of ways to roll a number in {1, 2, 3} is obviously just the cardinality of {1, 2, 3}. Hence

$$\text{P(Roll is even given roll is in } \{1, 2, 3\}) = \frac{|\,\text{Even} \cap \{1, 2, 3\}\,|}{|\,\{1, 2, 3\}\,|} = \frac{1}{3}.$$

Since the ways to roll an even number in {1, 2, 3} is just P(Even $\cap$ {1, 2, 3}), and since the number of ways to roll in {1, 2, 3} is just P({1, 2, 3}), we can express our result in terms of probabilities alone:

$$\text{P(Roll is even given roll is in } \{1, 2, 3\}) = \frac{\text{P(Even} \cap \{1, 2, 3\}\text{)}}{\text{P(}\{1, 2, 3\}\text{)}} = \frac{1}{3}.$$

5.2 The Definition of Conditional Probability

Conditional Probability. We use a special notation to indicate the probability of one event given another. The probability of event H given event E is P(H | E). For this probability to be meaningful, the probability of E has to be greater than 0. The probability of event H given the null event E = {} is undefined. The probability P(H | E) is called a *conditional probability.* So, the probability of rolling an even number given a roll in {1, 2, 3} is

P(Roll is even | Roll is in {1, 2, 3}).

Our reasoning for calculating this probability does not depend on any details involving dice; we generalize. We thus obtain the following result:

$$P(H \text{ given } E) = P(H \mid E) = \frac{P(H \cap E)}{P(E)} \text{ with } P(E) > 0.$$

Now we can see why P(E) must be greater than 0 – division by 0 is undefined. And this is an expression of the fact that it makes no sense to talk about looking for an event within the null event. You can't find anything in the empty set.

As another example, what is the probability of rolling an even number given that you've rolled a number in {4, 5, 6}? You can see just by counting that 2 out of 3 numbers in {4, 5, 6} are even, and since all are equally likely, the probability is 2/3. Let's work it out:

$H = \{2, 4, 6\};$

$E = \{4, 5, 6\};$

$$P(H \mid E) = \frac{P(\{2, 4, 6\} \cap \{4, 5, 6\})}{P(\{4, 5, 6\})} = \frac{2}{3}.$$

5.3 An Example Involving Marbles

When we compute the probability of H given E, we are restricting the sample space to just those cases in which E occurs. An example can help make this clearer. Consider a bag B that contains 700 marbles. Some are red, some are blue; some are opaque, some are translucent. The numbers of marbles in each category are given in Table 5.1.

	Opaque	**Translucent**	**Totals**
Red	180	120	300 total red
Blue	320	80	400 total blue
Totals	500 total opaque	200 total translucent	700 total marbles

Table 5.1 Some marbles and their attributes in bag B.

The experiment, in this example, is the blind selection of a marble: keeping his or her eyes closed, a person reaches his or her hand into bag B and pulls out a marble. We are interested in various conditional probabilities. For example, what is the probability that an opaque marble picked from B is red? In other words, that it is red given that it is opaque? We write this as P(Red in B | Opaque in B).

To figure this out, we want to ignore all the translucent marbles in B and focus only on the opaque marbles in B. One way to ignore all the translucent marbles and focus just on the opaque ones is to sort the marbles into two new bags. You go through all the marbles in the original bag. If a marble is opaque, you put it into the new bag O; if it is translucent, you put it into the new bag T. After doing that, you know that the probability P(Red in B | Opaque in B) is equal to the probability of picking a red marble from the opaque bag O. The opaque bag O contains 180 red opaque marbles and 320 blue opaque marbles. Thus

$$\text{P(Red in B} \mid \text{Opaque in B)} = \text{P(Red in O)} = \frac{|\text{Red in O}|}{|\text{Marble in O}|} = \frac{180}{500} = \frac{9}{25}.$$

Making up the two new bags O and T is both tedious and unnecessary. After all, these bags are already conceptually defined in Table 5.1. Any marble that is opaque in bag O was opaque when it was in bag B. And any marble that is red in bag O was both red and opaque in bag B. So, turning our attention back to the original bag B, we have

$$\text{P(Red in B} \mid \text{Opaque in B)} = \frac{\text{P(Red in B} \cap \text{Opaque in B)}}{\text{P(Opaque in B)}}.$$

And with a little calculation, we get these results:

$$\text{P(Red in B} \cap \text{Opaque in B)} = \frac{|\text{Red in B} \cap \text{Opaque in B}|}{|\text{Marble in B}|} = \frac{180}{700} = \frac{9}{35}.$$

$$\text{P(Opaque in B)} = \frac{|\text{Opaque in B}|}{|\text{Marble in B}|} = \frac{500}{700} = \frac{5}{7}.$$

Finally, a little algebra shows that

$$\text{P(Red in B} \mid \text{Opaque in B)} = \frac{9/35}{5/7} = \frac{9}{35} \cdot \frac{7}{5} = \frac{9 \cdot 7}{7 \cdot 5 \cdot 5} = \frac{9}{25}.$$

5.4 Independent Events

Independent Events. Suppose you roll a die and then roll it again. Nothing in the first roll influences the second roll (and vice versa). The two rolls are *independent*. Suppose F is the probability of rolling an even number on the first roll. This probability is P(F). Suppose E is the event of rolling an even number on the second roll. The probability of doing this is P(E). Since rolling an even number on the first roll has no influence on rolling an even number on the second roll, the probability of E given F does not differ from the probability of E itself. In symbols, P(E | F) = P(E).

We know that $P(E \mid F) = P(E \cap F) / P(F)$. And by multiplying both sides of that equation by $P(F)$, we get

$$P(E \cap F) = P(E \mid F) \cdot P(F).$$

If the events E and F are independent, then we can substitute $P(E)$ for $P(E \mid F)$ to get

$$P(E \cap F) = P(E) \cdot P(F).$$

Indeed, we can use this equation as a formal definition of independence. Thus E and F *are independent events* iff $P(E \cap F) = P(E) \cdot P(F)$.

6. Bayes Theorem

6.1 The First Form of Bayes Theorem

One of the most philosophically powerful results from the theory of probability is known as Bayes Theorem. Bayes Theorem is used extensively in epistemology and philosophy of science. Suppose H is some hypothesis and E is some evidence. Using our definition of conditional probability, we can express the probability of H given E as:

$$P(H \mid E) = \frac{P(H \cap E)}{P(E)} \text{ with } P(E) > 0.$$

But we can also reverse the conditions:

$$P(E \mid H) = \frac{P(H \cap E)}{P(H)} \text{ with } P(H) > 0.$$

So we can set $P(H \cap E) = P(E \mid H) \cdot P(H)$. Substituting into the first equation in this section, we get the first form of Bayes Theorem:

$$P(H \mid E) = \frac{P(E \mid H) \cdot P(H)}{P(E)} \text{ with } P(E) > 0.$$

The derivation of the first form of Bayes Theorem is simple. But the uses of this theorem are philosophically profound.

6.2 An Example Involving Medical Diagnosis

A clinic keeps precise records of the people who come for treatment. The clinic tracks both the symptoms of its patients and their diagnoses. Rcmarkably, this clinic has only ever seen two diseases: colds and allergies. And only two symptoms: coughs and sneezes. Equally remarkably, the symptoms never overlap. No one has ever shown up both sneezing and coughing. Nor has anyone ever shown up who has both a cold and allergies. It's all neat and tidy here. After a patient shows up, a simple blood test for histamine tells the doctors whether or not a person has allergies. The test is always perfectly accurate. It's a perfect world! The records kept by the clinic are in Table 5.2.

	Sneezing	**Coughing**	**Totals**
Allergies	200	800	1000 total with allergies
Cold	300	2700	3000 total with colds
Totals	500 total sneezers	3500 total coughers	4000 total patients

Table 5.2 Symptoms and diseases at a clinic.

One fine day, patient Bob walks up to the clinic. Before he even gets inside, nurse Betty sees him. She can't tell whether he is coughing or sneezing. But from past experience, before he even gets into the clinic, she knows that

P(Sneezing) = 500 / 4000 = 1/8;

P(Coughing) = 3500 / 4000 = 7/8.

So she knows that the odds are very high that he's coughing. And before patient Bob gets to the clinic, nurse Betty also knows something about the probabilities of his diagnosis. She knows that

P(Allergies) = 1000 / 4000 = 1/4;

P(Cold) = 3000 / 4000 = 3/4.

Prior Probabilities. Nurse Betty knows from past experience that it's quite likely that Bob has a cold. Since these are known before Bob even walks into the clinic, that is, before he ever reports any symptom or is diagnosed with any disease, these are the *prior probabilities*. Of course, the prior probabilities in this case aren't based on any facts about Bob at this time. They're based on past observations about people at other times (which may include Bob at some past time). These probabilities are *estimates*, based on past experience, that Bob has a symptom or a disease here and now. Based on past experience, it is reasonable to apply them to Bob. But they may have to be changed based on Bob's symptoms, and, ultimately, on the perfect diagnostic blood test.

The statistics kept by the clinic also allow us to compute some *prior conditional probabilities*. The probabilities that a patient has a symptom given that he or she has a disease are:

$$P(\text{Sneezing} \mid \text{Cold}) = 300 / 3000 = 1/10;$$

$$P(\text{Sneezing} \mid \text{Allergies}) = 200 / 1000 = 2/10;$$

$$P(\text{Coughing} \mid \text{Cold}) = 2700 / 3000 = 9/10;$$

$$P(\text{Coughing} \mid \text{Allergies}) = 800 / 1000 = 8/10.$$

Shortly after arriving, patient Bob is in an examination room. And now he's sneezing like crazy. Doctor Sue uses the clinic's past records and Bayes Theorem to quickly compute the probability that Bob has a cold given that he's sneezing:

$$P(\text{Cold} \mid \text{Sneezing}) = \frac{P(\text{Sneezing} \mid \text{Cold}) \cdot P(\text{Cold})}{P(\text{Sneezing})} = \frac{1/10 \cdot 3/4}{1/8} = \frac{3/40}{5/40} = \frac{3}{5}.$$

Posterior Probabilities. Given that Bob is sneezing, and given the database of past diagnoses compiled by the clinic, the probability that Bob has a cold is 3/5. Since this is the probability that he has a cold *after* he has manifested a symptom, that is, *after* he provides some evidence about his condition, this is the *posterior probability* that Bob has a cold. Since the prior probability that Bob has a cold is 3/4, and the posterior probability is only 3/5, the fact that Bob is sneezing decreases the probability that he has a cold.

You can see the relevance of Bayes Theorem to epistemology. Suppose our hypothesis is that Bob has a cold. Before Bob even walks in, the probability we assign to this hypothesis is the prior probability 3/4. When he sneezes, he provides us with some evidence. After providing that evidence, the probability that he has a cold goes down to 3/5. The use of Bayes Theorem shows how the evidence changes the probability of the hypothesis.

6.3 The Second Form of Bayes Theorem

We can use our analysis of conditional probability to derive a more general form of Bayes Theorem. We refer to it as the second form of Bayes Theorem. The second form is easier to apply in many situations.

To work through the derivation of the second form of Bayes Theorem, let's return to our example of the bag of marbles B from Table 5.1. Our hypothesis H is the event that a marble randomly taken from B is red. Our evidence E is the event that the marble is opaque. Of course, an opaque marble in B is either red or not red (i.e., it is blue). Hence

x is opaque iff ((x is opaque & x is red) or (x is opaque & x is not red)).

Letting –Red be the complement of Red (i.e., the set of marbles that are not red), we can express this in set-theoretic terms as

$$\text{Opaque} = ((\text{Opaque} \cap \text{Red}) \cup (\text{Opaque} \cap -\text{Red})).$$

Any marble is either red or not red; it can't be both. Hence the sets (Opaque $\cap$ Red) and (Opaque $\cap$ –Red) are mutually exclusive. Their intersection is empty. Since these two events are mutually exclusive, we know from our rule on unions of events that

$$P(\text{Opaque}) = P(\text{Opaque} \cap \text{Red}) + P(\text{Opaque} \cap -\text{Red}).$$

Since the event that a marble is opaque is our evidence E, and the event that it is red is our hypothesis H, we can write the equation above as

$$P(E) = P(E \cap H) + P(E \cap -H).$$

And with this we have reached an important philosophical point: the event –H can be thought of as a *competing alternative hypothesis* to H. The event H is the event that the marble is red; the competing hypothesis –H is the event that it is not red. A little algebra will help us express this in a very useful form.

Since $P(E \mid H) = P(E \cap H) / P(H)$, we know that

$$P(E \cap H) = P(H) \cdot P(E \mid H).$$

Since $P(E \mid -H) = P(E \cap -H) / P(-H)$, we know that

$$P(E \cap -H) = P(-H) \cdot P(E \mid -H).$$

And by substituting these versions of the intersections back in the earlier formula for P(E), we have

$$P(E) = P(H) \cdot P(E \mid H) + P(-H) \cdot P(E \mid -H).$$

And substituting this formula for P(E) into our first form of Bayes Theorem, we get our second form of Bayes Theorem:

$$P(H \mid E) = \frac{P(E \mid H) \cdot P(H)}{P(H) \cdot P(E \mid H) + P(-H) \cdot P(E \mid -H)}.$$

6.4 An Example Involving Envelopes with Prizes

We can use a selection game to illustrate the second form of Bayes Theorem. The game involves two large bags filled with ordinary letter-sized envelopes. Some of these envelopes contain dollar bills while others are empty. You are told the proportions of prize envelopes in each bag. Specifically, bag Alpha has 20 prize envelopes and 80 empty envelopes while bag Beta has 60 prize envelopes and 40 empty ones. To play this game, you make two selections. First, you randomly choose a bag (you pick blindly – you don't know if you've picked Alpha or Beta). Second, you randomly choose an envelope from a bag. You open the envelope to see whether or not you've won a prize. On your first selection, you discover that indeed you've won. Congratulations! Now, what is the probability that you chose your winning envelope from bag Beta?

We can directly apply the second form of Bayes Theorem. We have two hypotheses: the first is that your envelope came from Beta while the second is that it came from Alpha. These are mutually exclusive – either your envelope came from one bag or else it came from the other, but it did not come from both. Since the two hypotheses are mutually exclusive, we let H be the hypothesis that your envelope came from bag Beta and –H be the hypothesis that it came from Alpha. The evidence is just that your envelope contains a dollar bill. We need to define four probabilities:

P(H) = ?
P(–H) = ?
P(E | H) = ?
P(E | –H) = ?.

The description of the game provides us with all the information we need to know to determine these probabilities. First of all, you had to choose one bag or the other. Hence P(H) + P(–H) = 1. And the choice of the bag was random. It is equally likely that you chose Alpha as that you chose Beta. So P(H) = P(–H). Therefore, you can substitute P(H) for P(–H) and obtain P(H) + P(H) = 1, from which it follows that P(H) = 1/2 = P(–H). We've now established that

P(H) = 1/2
P(–H) = 1/2
P(E | H) = ?
P(E | –H) = ?.

The conditional probability P(E | H) is the probability that you got a prize given that you selected an envelope from Beta. This is just the proportion of prize envelopes in Beta. Since there are 100 total envelopes in Beta, and of those 60 are prize envelopes, we conclude that P(E | H) = 60 / 100 = 6/10. Likewise P(E | –H) = 20/100 = 2/10. Hence

$$P(H) = 1/2$$
$$P(-H) = 1/2$$
$$P(E \mid H) = 6/10$$
$$P(E \mid -H) = 2/10.$$

Plugging all these results into the second form of Bayes Theorem, we get

$$P(H \mid E) = \frac{1/2 \cdot 6/10}{1/2 \cdot 6/10 + 1/2 \cdot 2/10} = \frac{3/10}{4/10} = \frac{3}{10} \cdot \frac{10}{4} = \frac{3}{4}.$$

We therefore conclude that, given that you picked a winning envelope, the probability that you picked it out of bag Beta is 3/4.

7. Degrees of Belief

7.1 Sets and Sentences

At this point, a probability function maps a set onto a number. But, with a little conceptual work, we can also define probability functions on sentences. Let's see how to do this. Consider the case of a bag filled with marbles as in Table 5.1. If we randomly select an object from the bag, we can say things like "The chosen object is a marble" or "The chosen object is red" or "The chosen object is translucent". Any sentence involving a marble from the bag involves some predicate. There are five predicates in Table 5.1: x is a marble; x is red; x is blue; x is opaque; and x is translucent.

Suppose we let X stand for the "the chosen object". For any predicate F, the sentence "X is F" is either true or false when an object is chosen. Hence for any sentence of the form "X is F", there is a set of outcomes at which it is true and a set at which it is false. We can think of the outcomes as possible worlds at which the sentence is true (or false). According to our semantic work in Chapter 4 sec. 2, the proposition expressed by a sentence is the set of all possible worlds at which it is true. We can take this to be the *extension* of the sentence. For the experiment of choosing marbles from the bag defined in Table 5.1, each possible world is the choice of a marble. Hence there are as many possible worlds as marbles in the bag. And each set of possible worlds is a set of marbles. Hence the extension of any sentence of the form "X is F" is a set of marbles. So

the extension of "X is F" = $\{ x \text{ is a marble in the bag} \mid x \text{ is F} \}$.

As long as we're working through an example involving a single fixed sample space, all variables can be assumed to range over outcomes in that sample space. So we don't need to make this assumption explicit. In our current example, all variables range over objects in the bag B. We can drop the reference to x being in the bag B and just write the extension of "X is F" = $\{ x \mid x \text{ is F} \}$.

All this leads directly to the concept of probabilities for sentences. On any given choice, the probability that "The chosen object is red" is the probability that the chosen object is in the set of red objects. And that probability is the number of the set of red objects divided by the number of objects in the bag. In our set-theoretic language, it is P(Red). That is, P("X is red") = P(Red). Since writing the quotes in "X is F" is both annoying and unnecessary when writing P("X is F"), we drop them. Generally, for any predicate F,

P(X is F) = P(F).

We can go on to define probabilities for conjunctions of sentences. Consider "X is red and X is translucent". This is just the probability that X is in the intersection of the set of red marbles with the set of translucent marbles. As a rule,

P(X is F & X is G) = $P(F \cap G)$.

Disjunctions follow naturally too. Consider "X is blue or X is translucent". This is just the probability that X is in the union of the set of blue marbles with the set of translucent marbles. As a rule,

P(X is F or X is G) = $P(F \cup G)$.

Negations are straightforward. Consider "X is not opaque". This is just the probability that X is in the complement of the set of opaque marbles. We denote the complement of set F relative to the assumed background sample space as –F. Hence

P(X is not F) = P(–F).

7.2 Subjective Probability Functions

Any linguistically competent human mind is able to consider sentences in some language (which might, without too much worry, be called its language of thought or mentalese). For example, you can affirm that "Socrates was a philosopher" and deny that "No man is mortal". For precision, let's define a language we'll call *Extensional English*. The vocabulary of Extensional English splits into constants and predicates. The constants are all the names (proper nouns) in ordinary English. The predicates are the nouns, adjectives, and verbs in ordinary English. The grammar of Extensional English is the grammar of the predicate calculus. Here are some sentences in Extensional English:

philosopher(Socrates);

(wise(Plato) and man(Aristotle));

(for all x)(if man(x) then mortal(x)).

An *extensional mind* is one whose language of thought is Extensional English. For any sentence in Extensional English, an extensional mind believes it more or less. It assigns some *degree of belief* to the sentence. Let's use EE to denote the set of sentences in Extensional English. An extensional mind has a *doxastic function* D that maps every sentence in EE onto a degree of belief. For any sentence s in EE, the degree of belief D(s) is a number that indicates how much the mind believes s.

There are many ways to define doxastic functions. We are especially interested in doxastic functions that satisfy certain rules relating to probability. These are the axioms of the probability calculus. A *subjective probability function* P is a doxastic function that satisfies these axioms. There are three axioms.

The first axiom says that degrees of belief vary continuously between 0 and 1. In symbols,

$0 \leq P(S) \leq 1$ for any sentence S in EE.

The second axiom says that sentences that are certainly false are given the minimum degree of belief while sentences that are certainly true are given the maximum degree of belief. Contradictions (like "It's Tuesday and it's not Tuesday") are certain falsehoods while tautologies (like "Either it's Tuesday or it's not Tuesday") are certain truths. Hence

P(S) = 0 if S is a contradiction in EE;

P(S) = 1 if S is a tautology in EE.

For example,

P(wise(Plato) or not wise(Plato)) = 1;

P(wise(Plato) and not wise(Plato)) = 0.

The third axiom says that if sentences G and H are mutually exclusive, so that G is equivalent to not H, then the probability of (G or H) is the probability of G plus the probability of H. We write it like this:

P(G or H) = P(G) + P(H) if G is equivalent to not H.

For example, nurse Betty has four sentences in her language of thought about patients at the clinic: "The patient is sneezing"; "The patient is coughing"; "The patient has a cold"; and "The patient has allergies". Her subjective probability function assigns a degree of belief to each of these sentences at any moment of time. As time goes by, and her experience changes, her subjective probability function changes too.

8. Bayesian Confirmation Theory

8.1 Confirmation and Disconfirmation

We spoke casually about evidence E and hypothesis H. And we've discussed cases in which E increases the probability of H. For example, in our example with the clinic, the prior probability that Bob has allergies was 1/4, and the posterior probability was 2/5. So the fact that Bob was sneezing *increased* the probability that he had allergies. The sneezing evidence confirmed the hypothesis that he had allergies. More formally,

evidence E *confirms* hypothesis H whenever $P(H \mid E) > P(H)$.

The degree to which E confirms H is proportional to the difference between $P(H|E)$ and $P(H)$. The simplest way to put this is to say that if E confirms H, then

the degree to which E confirms H = $(P(H|E) - P(H))$.

As is often the case, the simplest way may not be the best. There are other ways to measure the degree of confirmation based on comparing $P(H|E)$ with $P(H)$. But we need not go into those details here.

Evidence need not confirm a hypothesis. In our example with the clinic, the prior probability that Bob has a cold was 3/4, and the posterior probability was 3/5. So the fact that Bob was sneezing *decreased* the probability that he had a cold. The sneezing evidence disconfirmed (or undermined) the hypothesis that Bob had a cold. As a rule,

evidence E *disconfirms* hypothesis H whenever $P(H \mid E) < P(H)$.

And, as expected, if E disconfirms H, then the simplest way to define the degree of disconfirmation is to say that

the degree to which E disconfirms H = $(P(H) - P(H|E))$.

Only one alternative remains: the evidence E has no effect on the probability of the hypothesis. In this case, the evidence is neutral. That is,

evidence E *is neutral with respect to* H whenever $(P(H \mid E) = P(H))$.

8.2 Bayesian Conditionalization

As any survey of the recent literature will show, Bayes Theorem is heavily used in decision theory, epistemology and philosophy of science. Much has been written under the subject headings of *Bayesian confirmation theory, Bayesian decision theory,* and *Bayesian epistemology.* But to apply Bayes Theorem, we need some additional formal machinery. We need a rule that allows us to use Bayes Theorem to change our degrees of belief. For example, recalling our discussion of poor Bob in Section 6.2 above, once we learn that Bob is sneezing, we should change the degree to which we believe he has a cold. The principle of *Bayesian conditionalization* tells us how to make such changes.

We start with a prior subjective probability function P. Whenever we change a degree of belief, we are defining a new subjective probability function. We are defining a posterior function P^+. Our prior function P might assign a degree of belief to an evidential statement E. For instance, the case of Bob and the clinic, we have P(Bob is sneezing) = 1/8. But when we learn that Bob is sneezing, we have to change this to 1. This change results in a new posterior function P^+. That is, P^+(Bob is sneezing) = 1. But this isn't the only change in the probabilities we can assign to sentences. For given that Bob is sneezing, the probability that he has a cold changes from 3/4 to 3/5. If you believe that Bayes Theorem is a good way to adjust your beliefs about the world, then you'll change the degree to which you believe that Bob has a cold from 3/4 to 3/5. You'll set P^+(Bob has a cold) to P(Bob has a cold | Bob is sneezing). This is an example of Bayesian conditionalization.

The principle of Bayesian conditionalization says (roughly) that whenever your degree of belief in an evidentiary sentence E changes to 1, you should use Bayes Theorem to update every sentence that is confirmed or disconfirmed by this change. More precisely, for any hypothesis H that is confirmed or disconfirmed by the change in P(E), you should set $P^+(H)$ to $P(H|E)$. Smoothing out the roughness in this principle requires us to take into consideration the fact that (1) many evidentiary sentences can change at once and the fact that (2) the change in P(E) is rarely likely to go to exactly 1. After all, we said that only tautologies (logical truths) have probabilities of 1, and contingent facts like Bob sneezing are hardly logical truths. But for the most part, we can ignore these subtleties. The main idea is that Bayesian conditionalization is a way to use Bayes Theorem to update our beliefs. Table 5.3 shows this in the case of Bob at the clinic. If nurse Betty uses Bayesian conditionalization to update her subjective probability function, then she changes her degrees of belief to match the degrees in Table 5.3.

Prior Probability Function			Posterior Probability Function		
P(Bob is coughing)	=	7/8	P^+(Bob is coughing)	=	0
P(Bob is sneezing)	=	1/8	P^+(Bob is sneezing)	=	1
P(Bob has a cold)	=	3/4	P^+(Bob has a cold)	=	3/5
P(Bob has allergies)	=	1/4	P^+(Bob has allergies)	=	2/5

Table 5.3 Change in subjective probabilities.

9. Knowledge and the Flow of Information

We've talked about evidence without much consideration of what it means to be evidence. One popular idea is that evidence is provided by observation – you set P(Bob is sneezing) to 1 because you see and hear Bob sneezing. Some philosophers have thought that you can't really see Bob sneezing. You have an experience of a certain sort. Your experience may or may not represent something happening in the external world. If your experience is *veridical*, then it represents the fact that Bob is sneezing. As you already surely know, you might be wrong. Maybe you're being deceived by an Evil Demon. Or maybe you're in a computer-generated hallucination like the Matrix. Or you're a brain in a vat behind the veil of ignorance on Twin Earth. Or whatever. Under what conditions can we say that your perceptual experience represents the fact that Bob is sneezing?

One answer is given by using conditional probability. Following Dretske (1981: 65), we might say that an experience E *carries the information* that x is F or represents the fact that x is F iff the conditional probability that x is F given E is 1. More formally,

experience E *represents that* x is F iff $P(x \text{ is F} \mid E) = 1$.

This analysis ignores several qualifications about your mind and its relation to the world. To learn more about those qualifications, you should read Dretske (1981). The biggest problem with this analysis of representation is that it seems to work only for minds located in ideal worlds. After all, it's reasonable to think that representation is a *contingent* relation between our minds and the external environment. And since something with a probability of 1 is a logical truth, it follows that for any mind in a less than ideal world, $P(x \text{ is F} \mid E)$ is less than 1. Another way to say this is to say that there is always some noise in any communications channel. Hence no mind in our universe ever represents anything. You might try to work out the consequences of allowing $P(x \text{ is F} \mid E)$ to be less than 1.

One interesting feature of the conditional analysis of representation is that it does not require causality (see Dretske, 1981: 26-39). Information can flow from a sender to a receiver without any causal interaction. The sender need not cause any effect in the receiver in order to send a signal to the receiver. And the receiver can represent what is going on in the sender without having been affected in any way by the sender. All that is needed is that the conditional probability is satisfied. Consider two perfectly synchronized clocks. Since they are synchronized, they always show the same time. Hence

P(Clock 1 shows noon | Clock 2 shows noon) = 1.

Each perfectly represents what is going on at the other without any causal interaction. A signal is sent without a cause. Leibniz was fond of such examples, as they allow his monads to perceive without having any windows (see Rescher, 1991: 258).

6

UTILITARIANISM

1. Act Utilitarianism

1.1 Agents and Actions

Among ethical theories, utilitarian ones sometimes involve a fair amount of mathematics. This marriage of mathematics with morals goes back to the 19th century founders of utilitarianism, namely, Bentham and Mill. There are many versions of utilitarianism. One version – *act utilitarianism* – says that acts have a property known as their utility. The utility of an act determines whether it is right or wrong for an agent (a person) to perform the act. For example, Feldman says: "An act is right if and only if its utility is at least as great as that of any of its alternatives" (1997: 21). He likewise says an act is wrong iff it is not right and an act is obligatory iff it would be wrong to not do it (1997: 21).

Agent. Any version of act utilitarianism involves agents, actions, and consequences. For our purposes, an *agent* is a machine that is situated in some context. An agent M is a triple (P, K, n). The item P in an agent is the program of M. It is the essence of the agent. The item K is a career of the machine M. It is a series of configurations of M. Hence, K is a function from some number x to the set of configurations C_M of machine M. The item n is the individuating number of the agent. After all, two distinct agents may have the same program and career (recall dual universes or the eternal return). We distinguish agents with the same program and career by giving them distinct numbers. Agents are situated in worlds where they are logically – and ethically – related to other agents. A world is just a system of interacting agents. On this model, it is a network of machines.

Action. An *action* is a transition from one configuration of an agent to another. Any action is a member of the set of possible actions for the agent. These are determined by its state-transition network. For a machine M, a possible action of M is a pair (x, y) where x and y are configurations of M (that is, x and y are in C_M) and y is a successor of x. For any machine M, its set of possible actions is

$$\text{Acts}(\text{ M}) = \{ (x, y) \mid \text{x} \in C_M \ \& \ \text{y} \in C_M \ \& \ y \text{ is a successor of } x \}.$$

Any two agents with the same program have the same sets of possible actions. Let's work this out more precisely. Consider agent Alpha = (P, J, m) and agent Beta = (P, K, n). These two agents have the same program (the program P). Hence their *possible* actions are the same. But these two agents have different careers. Alpha has career J and Beta has career K. Hence, while their possible action sets are identical, their actual action sets are distinct. The actual actions of Alpha are the transitions in its career. The actual actions of Beta are the transitions in its career. The consequences of these actions involve other agents in the worlds of Alpha and Beta. You can suppose that these worlds are different. Suppose Alpha performs a certain action and Beta performs the same action. When Alpha performs it, the consequences may be pleasurable; when Beta performs it, they may be painful. It all depends on their worlds, that is, on their relations with other agents.

1.2 Actions and their Consequences

When an agent performs an act, it can have some pleasurable consequences. These are its *hedonic consequences*. A hedonic consequence of an act is any experience of pleasure caused (however directly or indirectly) by that act in any agent at any time. We could analyze the concept of a hedonic consequence in great detail. But right now we don't need to. Right now we say only that the hedonic consequences of act A are

$$\text{HC(A)} = \{ x \mid x \text{ is a pleasurable experience caused by act A} \}.$$

Following traditional utilitarianism, we say that any experience of pleasure has three features: its duration; its intensity; and its quality. For our purposes, all transitions take place in a single time-step of the world. So we ignore duration. The intensity of any hedonic consequence of A is its *hedonic intensity* (HI). The quality of any hedonic consequence of A is its *hedonic quality* (HQ). The *hedonic value* (HV) of any hedonic consequence is the product of its quality and intensity. Formally, if x is any hedonic consequence of A, then the hedonic value HV of x is

$$\text{HV}(x) = \text{HQ}(x) \cdot \text{HI}(x).$$

To obtain the *gross hedonic value* (GHV) of an act, we take the sum of the hedonic values of all its hedonic consequences. Roughly speaking, this is the sum of all the pleasures caused by the act. More precisely speaking, for any act A, it is

$$\text{GHV(A)} = \text{the sum, for all } x \text{ in HC(A), of HV}(x) = \sum_{x \in \text{HC(A)}} \text{HV}(x).$$

Our analysis of pain follows our analysis of pleasure. When an agent performs an act, it can have some painful consequences. Pain is said to be *doloric*. A *doloric consequence* (DC) of an act is any experience of pain caused by that act in any agent at any time. We say

$$\text{DC(A)} = \{ x \mid x \text{ is a painful experience caused by act A} \}.$$

Painful experiences, like pleasurable experiences, have intensities and qualities. The intensity of a doloric consequence is DI. The quality of a doloric consequence is DQ. The value DV of a doloric consequence x is

$$DV(x) = DQ(x) \cdot DI(x).$$

As with pleasures, to obtain the *gross doloric value* (GDV) of an act, we take the sum of the doloric values of all its doloric consequences. Roughly speaking, this is the sum of all the pains caused by the act. More precisely speaking, for any act A, it is

$$GDV(A) = \text{the sum, for all } x \text{ in } DC(A), \text{ of } DV(x) = \sum_{x \in DC(A)} DV(x).$$

1.3 Utility and Moral Quality

Many utilitarians have said that the only thing that makes anything good is the pleasure it involves or produces, and the only thing that makes anything bad is the pain it involves or produces. To put it roughly, pleasure is good and pain is evil. Accordingly, the utility of an act is the total pleasure it causes minus the total pain it causes. We symbolize the utility of an act A as U(A) and we define it like this:

$$U(A) = GHV(A) - GDV(A).$$

We said that an act is right for agent x at time t iff its utility is at least as great as the utility of any other alternative act. The alternatives to an act are just the other transitions with the same initial member as act A. Thus

$$ALT(A) = \{ (x, z) \mid x \text{ is the first item in A } \& \; z \text{ is a successor of } x \}.$$

Finally, our simple version of act utilitarianism says that an act A is right iff its utility is at least as great as the utility of every alternative act B:

$$\text{Right}(A) \text{ iff (for all B)(if B} \in ALT(A) \text{ then } U(A) \geq U(B)).$$

2. Expected Utility

You flip a coin over and over. The coin is fair. The probability of it coming up heads is 1/2, and the probability of it coming up tails is 1/2. If it comes up heads, you get 1 dollar; if it comes up tails, you get 0 dollars. Suppose you flip the coin many times. In the long run, how much should you expect to win? You should expect heads to come up in 1/2 of the times you toss the coin; and you get $1 for each head. So you should expect to get:

$$\text{expected value} = \text{probability of heads} \cdot \text{value of heads} \cdot \text{number of tosses.}$$

More formally, let Ex(A, *n*) be the expected value of performing the action A *n* times. The probability that the coin comes up heads given that it is tossed is P(Heads | Toss). The value of heads is its utility. We write this as U(Heads). The number of times you can expect to get heads is P(Heads | Toss) · *n*. Multiply that by the value you get each time the coin comes up heads. Thus

$$\text{Ex(Toss}, n) = (\text{P(Heads | Toss)} \cdot n) \cdot \text{U(Heads)}.$$

For example, if you toss the coin 100 times, you should expect the coin to come up heads 50 times. Since each time it comes up heads, you get 1 dollar, you should expect to get 50 dollars. Of course, what you expect to get and what you actually get might be different. Your expectation is an estimate in an uncertain situation.

Now suppose you can win something on tails as well. If the coin comes up tails, you'll get 50 cents. If you toss the coin *n* times, then your total payoff includes both the payoff for your heads as well as for your tails. It is (the expected number of times it comes up heads · the utility of heads) + (the expected number of times it comes up tails · the utility of tails). We can work it out formally like this:

$$\begin{aligned}\text{Ex(Toss}, n) = \; & ((\text{P(Heads | Toss)} \cdot n) \cdot \text{U(Heads)}) + \\ & ((\text{P(Tails | Toss)} \cdot n) \cdot \text{U(Tails)}).\end{aligned}$$

For example, if you toss the coin 100 times, you should expect to get 50 heads and 50 tails. Thus your payout is $(50 \cdot 1) + (50 \cdot 0.5) = 75$ dollars.

You might just toss the coin once. When we compute the expected value of a single toss, we are computing Ex(Toss, 1). Since the 1 doesn't affect the multiplication (any number times 1 is itself), we can equate the expected value Ex(Toss) with Ex(Toss, 1). The equation for this is

$$\text{Ex(Toss)} = (\text{P(Heads | Toss)} \cdot \text{U(Heads)}) + (\text{P(Tails | Toss)} \cdot \text{U(Tails)}).$$

We can generalize this logic to determine the expected value or expected utility of any action. The action has some consequences. For instance, the action is tossing a coin. Let this action be A. The consequences are heads or tails. The consequences are C_1 or C_2. And, of course, they are mutually exclusive. You can't get both heads and tails on the same toss. We write our more general equation as

$$\text{Ex(A)} = (\text{P}(C_1 \mid \text{A}) \cdot \text{U}(C_1)) + (\text{P}(C_2 \mid \text{A}) \cdot \text{U}(C_2)).$$

The equation applies regardless of the nature of the action or consequences. That is, in the equation, A could be any action with any two mutually exclusive outcomes C_1 and C_2. We can generalize to any number of outcomes. Suppose you roll a six-sided die. Each side is associated with a payoff – with a utility. Hence the expected utility is:

$$\text{Ex(Roll)} = \sum_{i=1}^{6} (\text{P(Die shows } i \mid \text{Roll)} \cdot \text{U(Die shows } i)).$$

You can use the notion of expected utility to decide which action you should perform. Suppose you can perform any action in some set. Each action A is associated with a set of consequences. The cardinality of the set of these consequences is #A. The i-th consequence of doing A is $C_i(A)$. The expected utility of doing A is:

$$\text{Ex(A)} = \sum_{i=1}^{\#A} (\text{P(} C_i(\text{A}) \mid \text{A)} \cdot \text{U(} C_i(\text{A}))).$$

In the ideal case, you know the set of possible actions, the (mutually exclusive) consequences of each action, the probability of each consequence given the action, as well as the utility of each consequence. So you can calculate the expected utility of each possible action. Suppose one action stands out as providing a much greater expected utility than every other action. Intuitively, it would be prudent to do that action and foolish to do any other action. Somewhat more precisely, the notion of expected utility gives you a reason for doing the action with the greatest expected utility. Given some set of actions, it is *rational* to perform the action that produces the greatest expected utility. Thus expected utility enters into a plausible definition of rationality.

3. World Utilitarianism

3.1 Agents and their Careers

Act utilitarianism doesn't use much math. But the *world utilitarianism* of Feldman (1997) is highly mathematical. So we'll use it to illustrate the use of mathematics in ethics. As with our other examples, we are neither advocating nor criticizing Feldman's theory. Nor are we entirely faithful to Feldman's original presentation. World utilitarianism is a great opportunity to use mathematics to precisely develop a philosophical theory. We go into considerable technical detail. But it's worth it for two reasons. First, world utilitarianism illustrates many valuable concepts. It illustrates the use of branching times. These are relevant to discussions about possibility, free will, and related notions. It can be coupled with probability and expected utility. Second, our discussion of world utilitarianism gives you a good opportunity to hone your formal skills.

We start with a machine M. Machine M has various configurations and transitions. It would be nice if we could give an example in which M had trillions upon trillions of configurations and transitions. Then M would be more like a human person. But we can't. In our weakness, sadly, we can only muster a machine with 15 configurations. We use capital letters to denote them. So the set of configurations of machine M is

$$C_M = \{A, B, C, D, E, F, G, H, I, J, K, L, M, N, O\}.$$

Our notation is slightly ambiguous – the letter M denotes one of the configurations of the machine M. But we need not worry: context will always distinguish the two uses of the same letter. The configurations of machine M are linked by a successor relation. The successor relation defines the transitions from configuration to configuration. The configurations and the successor relation form a tree of branching careers of M. Figure 6.1 shows the tree. An arrow from one configuration to another is a possible transition of M. The right side of Figure 6.1 lists the career obtained by following branches to the end. The careers of M are in the set H_M. The eight careers of M are

$$H_M = \{ ABDH, ABDI, ABEJ, ABEK, ACFL, ACFM, ACGN, ACGO\}.$$

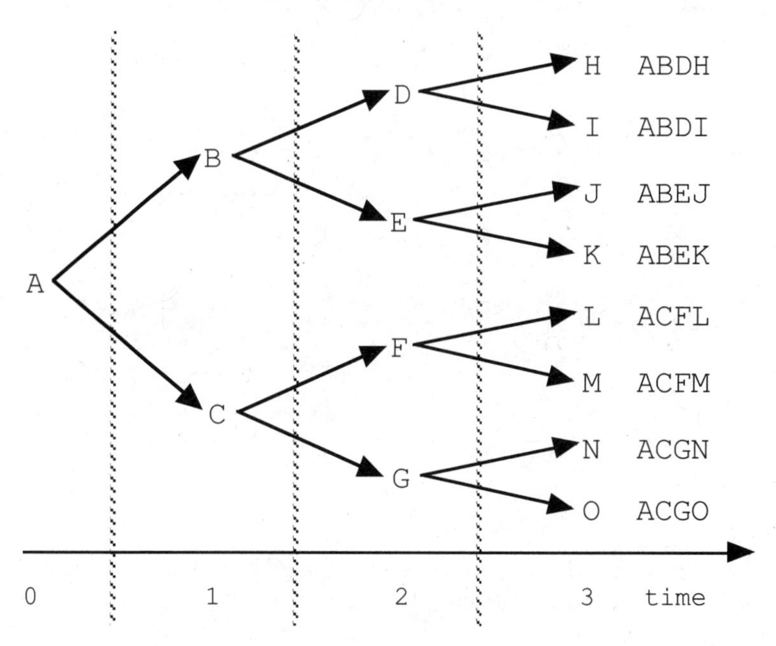

Figure 6.1 The branching transitions of machine M.

3.2 Careers and Compatible Worlds

A career is (or is not) a part of a possible world. A world is *compatible* with a career iff that career is part of that world. We think of worlds as mechanical universes – as closed networks of interacting machines. For this simple example, we allow worlds to overlap. Worlds can share parts. Thus the configuration A may be in many worlds and the career ABDH may be a part of many worlds. As usual, we neither mean to endorse nor to criticize this conception of worlds. Perhaps, as David Lewis argues, worlds don't overlap. But for now, for any career, there is a set of worlds with which it is compatible:

$$\text{COMP}(h) = \{ w \mid \text{career } h \text{ is a part of world } w \}.$$

We need to supply every career of M with some compatible worlds. We do this in Table 6.1. This table associates every career of M with one or more compatible worlds. Note that a career can be compatible with many worlds. This is because the career of M does not fully determine any of these worlds. The worlds contain events that are independent of the career of M. For example, things not involving M happen differently at w_1 and w_2.

Career	**Compatible Worlds**
ABDH	w_1, w_2
ABDI	w_3, w_4
ABEJ	w_5
ABEK	w_6, w_7, w_8
ACFL	w_9
ACFM	$w_{10}, w_{11}, w_{12}, w_{13}$
ACGN	w_{14}, w_{15}
ACGO	w_{16}, w_{17}, w_{18}

Table 6.1 Careers and the worlds that contain them.

3.3 Runs and Prefixes

A *run* of a machine M is any sequence of configurations M whose members are linked by the successor relation. H is a run of M iff H is a sequence of configurations of M such that each next configuration is a successor of the previous configuration. More precisely, H is a run of M iff H is a sequence $\{H_0, \ldots H_n\}$ of configurations of M such that, for i varying from 0 to n-1, H_{i+1} is a successor of H_i.

It's easy to use the branching tree from Figure 6.1 to list the runs of M. Each run is just a path that starts from configuration A and ends somewhere else in the tree. A run need not go all the way to the end of the tree. For example, AB is a run, as is ACG. Of course, a career is a complete run; it is a run that goes to the end of the tree. For any machine M, the runs of M are in the set R_M. For our sample machine, the runs are

R_M = { A, AB, AC, ABD, ABE, ACF, ACG,
ABDH, ABDI, ABEJ, ABEK, ACFL, ACFM, ACGN, ACGO }.

One run can be a prefix of another run. We don't need to formalize this – the notion of a prefix is clear enough. When we talk about prefixes, we are including only *proper prefixes*. We do not regard any run as a prefix of itself. For example, the run A is a prefix of every run; the run AB is a prefix of ABD and ABDH. The run ABDH is not a prefix of any run. Table 6.2 shows the prefix relation for the runs of our sample machine M. In Table 6.2, run *h* is a prefix of run *k* iff *h* is to the left of *k*.

<table>
<tr><td rowspan="8">A</td><td rowspan="4">AB</td><td rowspan="2">ABD</td><td>ABDH</td></tr>
<tr><td>ABDI</td></tr>
<tr><td rowspan="2">ABE</td><td>ABEJ</td></tr>
<tr><td>ABEK</td></tr>
<tr><td rowspan="4">AC</td><td rowspan="2">ACF</td><td>ACFL</td></tr>
<tr><td>ACFM</td></tr>
<tr><td rowspan="2">ACG</td><td>ACGN</td></tr>
<tr><td>ACGO</td></tr>
</table>

Table 6.2 The prefixes of runs.

3.4 Runs and the Worlds Open to Them

At any time before the end of your life, you have many possibilities. You have many *open futures*. Focus on some time just before you go to college. You are thinking about your major. You might major in philosophy or you might major in mathematics or you might major in both (good for you!). All these are open futures for you at that time. If you might do something, there are many worlds in which you do it. Many worlds are compatible with your life at the time you're deciding on your major. There are worlds in which you major only in philosophy; worlds in which you major only in mathematics; and worlds in which you major in both. At that time, all those worlds are open to you.

As time goes by, fewer and fewer worlds are open to any agent. As we grow older, we lose possibilities. Let's illustrate this with an example. At time 1, Sue is a senior in college. She might go to law school, or she might go to medical school. The worlds in which she goes to law school are w_1 and w_2. The worlds in which she goes to medical school are w_3 and w_4. So, at time 1, the worlds open to Sue are {w_1, w_2, w_3, w_4}. From time 1 to time 2, Sue makes a decision. She decides to go to medical school, and she does. She won't be going to law school. So at time 2, it is no longer possible for Sue to go to law school. The law school worlds w_1 and w_2 are now *closed*. Those possibilities are *lost*. The worlds open to Sue at time 2 are w_3 and w_4. The world w_3 is one in which Sue becomes an internist. The world w_4 is one in which she becomes a surgeon. From time 2 to time 3, Sue decides to become a surgeon. And she does become one. World w_3 is thus closed to Sue. Only world w_4 remains open. Suppose we let OPEN(Sue, *t*) be the worlds open to Sue at time *t*. We thus have the following:

$$\text{OPEN(Sue, 1)} = \{w_1, w_2, w_3, w_4\};$$
$$\text{OPEN(Sue, 2)} = \{w_3, w_4\};$$
$$\text{OPEN(Sue, 3)} = \{w_4\}.$$

Since Sue, like all of us, is constantly shedding possibilities, each later set of open worlds is a subset of the earlier sets of open worlds. For example,

$$\text{OPEN(Sue, 1)} \supseteq \text{OPEN(Sue, 2)} \supseteq \text{OPEN(Sue, 3)}.$$

We can apply this reasoning to machines and their runs. For any run *h*, there are some worlds open to *h*. These worlds are available to *h*. They are the worlds that *h* could be in. Recall that a career is a complete run – it is an entire possible life of a machine. A world *w* is open to a run *h* iff there is some career *k* such that *h* is a prefix of *k* and *k* is compatible with *w*. For any run *h*, let OPEN(*h*) be the worlds open to *h*. Formally,

$$\text{OPEN}(h) = \{ w \mid \text{world } w \text{ is open to run } h \}.$$

For any run *h* in R_M, we can specify the set of worlds in OPEN(*h*). Table 6.3 specifies the open worlds for any run. You can think of time as moving from the left to the right in Table 6.3. The leftmost column is earliest. The rightmost column is latest. Any run is a prefix of all the runs to its right. You can see that the sets of open worlds shrink as you move from left to right (that is, from past to future).

<table>
<tr><td rowspan="8">A
$w_1 \ldots w_{18}$</td><td rowspan="4">AB
$w_1 \ldots w_8$</td><td rowspan="2">ABD
$w_1 \ldots w_4$</td><td>ABDH
w_1, w_2</td></tr>
<tr><td>ABDI
w_3, w_4</td></tr>
<tr><td rowspan="2">ABE
$w_5 \ldots w_8$</td><td>ABEJ
w_5</td></tr>
<tr><td>ABEK
w_6, w_7, w_8</td></tr>
<tr><td rowspan="4">AC
$w_9 \ldots w_{18}$</td><td rowspan="2">ACF
$w_9 \ldots w_{13}$</td><td>ACFL
w_9</td></tr>
<tr><td>ACFM
$w_{10}, w_{11}, w_{12}, w_{13}$</td></tr>
<tr><td rowspan="2">ACG
$w_{14} \ldots w_{18}$</td><td>ACGN
w_{14}, w_{15}</td></tr>
<tr><td>ACGO
w_{16}, w_{17}, w_{18}</td></tr>
</table>

Table 6.3 Open worlds for runs.

3.5 The Utilities of Worlds

Every world has some utility. To compute this utility, we start with the idea that a world is a set of careers of machines. Each configuration of each career has some associated hedonic value HV and doloric value DV. As with agent utilitarianism, we might define these as products of intensity and quality. But we don't need to bother with those details a second time – there's no new math. We'll just get the hedonic value of a career by summing the hedonic values of its stages. So let STAGES(h) be the stages of career h. These are the configurations in h. Thus

$$\text{HV}(h) = \sum_{x \in \text{STAGES}(h)} \text{HV}(x).$$

To get the gross hedonic value of the whole world, we sum the hedonic values of all the careers in the world. It's instructive to show this as a double sum:

$$\text{GHV}(w) = \sum_{h \in w} \left[\sum_{x \in \text{STAGES}(h)} \text{HV}(x) \right].$$

We perform analogous calculations to obtain the doloric value of a career and the gross doloric value GDV of a world. The utility of a world is symbolized as UTIL(w) and is defined as the difference between the gross hedonic and doloric values:

$$\text{UTIL}(w) = \text{GHV}(w) - \text{GDV}(w).$$

No doubt there are many ways to compute the utility of a world. We might take averages instead of sums. The *average hedonic value* (AHV) of world w is the gross hedonic value divided by the number of careers in the world. Taking world w to just be a set of careers, AHV(w) is just GHV(w) divided by the cardinality of w:

$$\text{AHV}(w) = \text{GHV}(w) / |w|.$$

We can likewise compute the average doloric value and take the utility as their difference:

$$\text{UTIL}(w) = \text{AHV}(w) - \text{ADV}(w).$$

And surely there are other ways. But we don't need to worry about them here.

For our examples, we need some assignment of utilities to worlds. There's no point in spelling out the full contents of each world and calculating the utilities. You'd only be practicing sums, and you already know how to do sums. So we'll just give two sample assignments of utilities to worlds. These are Sample Utility Table 6.1 and Sample Utility Table 6.2. There is one best world in Sample Utility Table 6.1. It is world w_7. This world has greater utility than every other world. There are many equally maximally good worlds in Sample Utility Table 6.2. These are w_5, w_7, w_{10}, w_{13}, and w_{16}.

World	1	2	3	4	5	6	7	8	9
Utility	5	2	4	1	10	6	12	3	7

World	10	11	12	13	14	15	16	17	18
Utility	9	2	3	10	1	4	9	8	7

Sample Utility Table 6.1 There is one best world.

World	1	2	3	4	5	6	7	8	9
Utility	5	2	4	1	10	6	10	3	7

World	10	11	12	13	14	15	16	17	18
Utility	10	2	3	10	1	4	10	8	7

Sample Utility Table 6.2 There are many maximally good worlds.

3.6 Optimal Open Worlds

For any run h, there is a set of *optimal open worlds*. An optimal open world for h is a world that is open for h and that is at least as good as any other world that is open for h. Of course, goodness here is just utility. So an optimal open world for *h* is one whose utility is greater than or equal to that of any other open world for *h*. The first sketch of a symbolization looks like this:

> *x is an optimal open world for h* iff
> *x* is an open world for *h* &
> for any *y*, if *y* is open for *h*, then $\text{UTIL}(y) \leq \text{UTIL}(x)$.

The set of optimal open worlds for *h* is symbolized like this:

> $\text{OPT}(h) = \{ w \mid w \text{ is an optimal open world for } h \}$.

The distribution of optimal open worlds given Sample Utility Table 6.1 is shown in Optimal Open Table 6.1. The distribution of optimal open worlds given Sample Utility Table 6.2 is shown in Optimal Open Table 6.2.

It has to be stressed that the optimal open world for *h* is relative to *h*. For example, consider a case based on Sample Utility Table 6.1. At time 0, the optimal open world for machine M is w_7. This is the best of all possible worlds in Sample Utility Table 6.1. The run A can be extended in two ways: our machine can either make a transition to configuration B or to C. If it goes to B, then its run so far becomes AB. The change from A to AB keeps w_7 in view. It remains open. The change from A to AC does not keep w_7 in view. The best of all possible worlds is no longer an open world. The change from A to AC is an action that excludes w_7 from the set of open worlds. So now the best world that is available to AC is world w_{13}. Although w_{13} is not the best of all possible worlds, it is the best world that is possible for AC. It is the best world that is open to AC. If AC changes to ACF, it keeps the best world open; if it changes to ACG, it loses the best world available to it.

<table>
<tr><td rowspan="8">A
w_7</td><td rowspan="4">AB
w_7</td><td rowspan="2">ABD
w_1</td><td>ABDH
w_1</td></tr>
<tr><td>ABDI
w_3</td></tr>
<tr><td rowspan="2">ABE
w_7</td><td>ABEJ
w_5</td></tr>
<tr><td>ABEK
w_7</td></tr>
<tr><td rowspan="4">AC
w_{13}</td><td rowspan="2">ACF
w_{13}</td><td>ACFL
w_9</td></tr>
<tr><td>ACFM
w_{13}</td></tr>
<tr><td rowspan="2">ACG
w_{16}</td><td>ACGN
w_{15}</td></tr>
<tr><td>ACGO
w_{16}</td></tr>
</table>

Optimal Open Table 6.1 From Sample Utility Table 6.1.

<table>
<tr><td rowspan="8">A
$w_5, w_7, w_{10}, w_{13}, w_{16}$</td><td rowspan="4">AB
w_5, w_7</td><td rowspan="2">ABD
w_1</td><td>ABDH
w_1</td></tr>
<tr><td>ABDI
w_3</td></tr>
<tr><td rowspan="2">ABE
w_5, w_7</td><td>ABEJ
w_5</td></tr>
<tr><td>ABEK
w_7</td></tr>
<tr><td rowspan="4">AC
w_{10}, w_{13}, w_{16}</td><td rowspan="2">ACF
w_{10}, w_{13}</td><td>ACFL
w_9</td></tr>
<tr><td>ACFM
w_{10}, w_{13}</td></tr>
<tr><td rowspan="2">ACG
w_{16}</td><td>ACGN
w_{15}</td></tr>
<tr><td>ACGO
w_{16}</td></tr>
</table>

Optimal Open Table 6.2 From Sample Utility Table 6.2.

3.7 Actions and their Moral Qualities

When a machine makes a transition, it *extends* its run. Run *k* is an *extension* of run *h* iff *h* is a prefix of *k*, *k* is one configuration longer than *h* and the last configuration of *k* is a successor of the last configuration of *h*. For example, AB is an extension of A; ABD is an extension of AB; ABDH is an extension of ABD.

We have already characterized an action as a pair of configurations. But we can just as well characterize an action as a pair of runs. A pair of runs (*h*, *k*) is an action iff *k* is an extension of *h*. It follows that the set of possible actions for *h* is

$$\text{ACTS}(h) = \{ (h, k) \mid k \text{ is an extension of } h \}.$$

When *h* becomes *k*, then optimal open worlds are either saved or lost; they are kept open or closed off. According to world utilitarianism, saving or losing optimal open worlds is what determines the moral quality (right or wrong) of an action.

On the one hand, some of the optimal open worlds for *h* are *saved* in the transition to *k* (that is, as *h* is extended to *k*) iff the intersection of OPT(*h*) with OPT(*k*) is not empty. The saved optimal open worlds are in the intersection of OPT(*h*) with OPT(*k*). Since worlds are always being shed in the transition from *h* to *k*, it follows that if the intersection of OPT(*h*) with OPT(*k*) is not empty, then OPT(*k*) is a subset of OPT(*h*). If some of the optimal open worlds for *h* are saved in the transition to k, then h extended itself in one of the best available ways. It kept some optimal future open. It did the best it could in the present circumstances – where the best involves only the utilities of entire worlds. If you do the best you can, then, according to world utilitarianism, what you do is right. Thus an extension from h to k is *right* iff OPT(k) $\subseteq$ OPT(h). In terms of actions,

action (h, k) is *right* iff (OPT(k) $\subseteq$ OPT(h)).

For example, Right Sequence Table 6.1 shows a sequence of four right actions based on Sample Utility Table 6.2. For each right action, the later set of optimal open worlds is a subset of the earlier set of optimal open worlds. Every action in ABEK is right.

Run	A	AB	ABE	ABEK
Optimal Open Worlds	w_5, w_7, w_{10}, w_{13}, w_{16}	w_5, w_7	w_5, w_7	w_7

Right Sequence Table 6.1 A sequence of morally right actions.

On the other hand, none of the optimal open worlds for h at t are saved iff OPT(h) is not a subset of OPT(k). If none of the optimal open worlds are saved, then all are lost. That is, the best available alternatives for h are lost and h did not act optimally (the machine M did not extend itself optimally). It did not keep any best future in view; it did not do the best it could do in the circumstances. Hence the action of h is wrong. More formally,

action (h, k) is *wrong* iff (OPT(h) $\cap$ OPT(k) = {}).

Given Sample Utility Table 6.1, we can assign rightness and wrongness to all the transitions in the tree of M. Right is R and wrong is W. The assignment is shown in Moral Assignment 6.1. Given Sample Utility Table 6.2, we can assign rightness and wrongness to all the transitions in the tree of M. This is Moral Assignment 6.2.

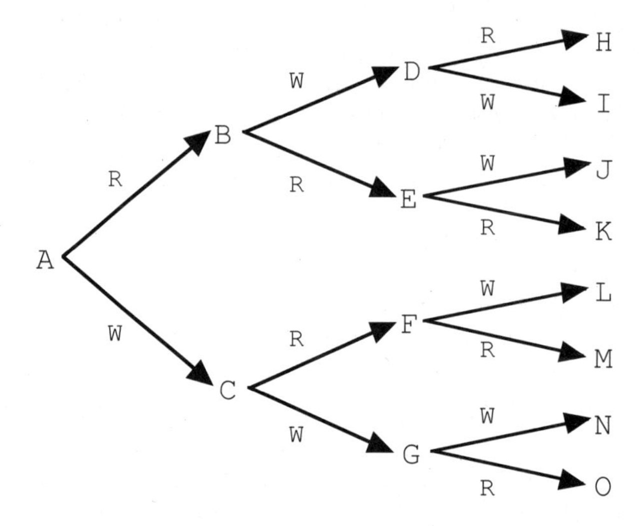

Moral Assignment 6.1 Based on Sample Utility Table 6.1.

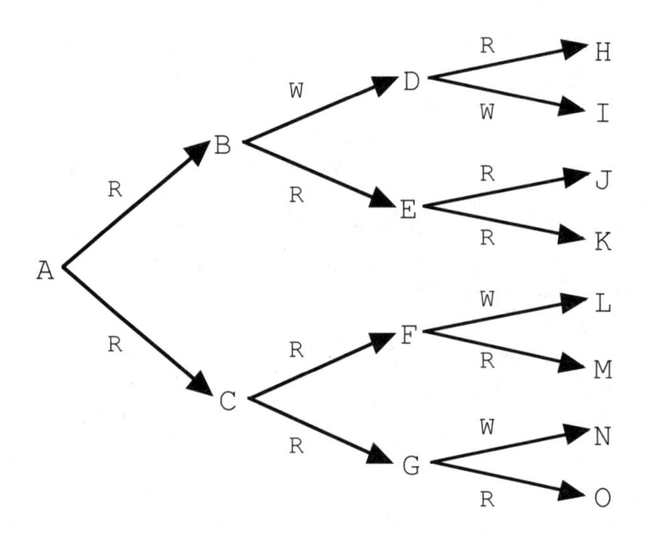

Moral Assignment 6.2 Based on Sample Utility Table 6.2.

7

FROM THE FINITE TO THE INFINITE

1. Recursively Defined Series

Some relations are defined *recursively*. The most basic kind of recursive definition has two parts: a basis clause and a recursion clause. We prefer to say that the basis clause is an *initial rule*, and the recursion clause is a *successor rule*.

For example, the relation is-a-descendent-of is defined recursively. The idea is that a child is a descendent; a child of a child is a descendent; a child of a child of a child is a descendent; and so on. We express this endless iteration of kids by saying that a child is a descendent and a child of a descendent is a descendent. The two rules are:

> *Initial Rule*. Every child of y is a descendent of y.
>
> *Successor Rule*. For any x, if x is a descendent of y, then every child of x is a descendent of y.

The kind *stroke series* is recursively defined. Informally, a stroke series is just a sequence of written strokes. Thus | is a stroke series; || is a stroke series; ||| is a stroke series; and so on. Here are the formal rules:

> *Initial Rule*. There exists an initial stroke series |.
>
> *Successor Rule*. For every x, if x is a stroke series, then x| is a stroke series. The stroke series x| is the successor of x.

Although a stroke series is something that a person can write down, the definition says nothing about people or the act of writing. It does not say that a stroke series has to be made or constructed by a person. The definition of a stroke series does not depend on persons or their constructive activities. It is entirely possible that there are natural stroke series that are not made by people (or by any other agents).

As you might expect, the sequence of natural numbers is recursively defined:

> *Initial Rule*. There exists an initial number 0.
>
> *Successor Rule*. For every number n, there exists a number $n+1$. The number $n+1$ is the successor of n and is greater than n.

Since these two rules define all the natural numbers, we can use them to define the set of all natural numbers (the extension of the property is-a-natural-number). According to convention, the set of natural numbers is N. N is defined like this:

Initial Rule. The initial natural number 0 is in N.

Successor Rule. For every x, if x is in N, then the successor of x is in N.

Putting these two rules together, we define N in a single sentence like this: N is the set of natural numbers iff 0 is in N and, for all x, if x is in N then $x+1$ is in N. And of course we need to add that there are no other objects in N.

A more directly philosophical example involves the organization of beliefs. Any mind has some beliefs. For convenience, we'll say it's *your* mind. Your beliefs are stratified into levels based on justification. Beliefs on higher levels are justified by beliefs on lower levels. For example, if P and P $\Rightarrow$ Q are on level 1, then Q is on level 2. Besides the initial and successor rules, we add a final rule that accumulates all your beliefs. The rules are:

Initial Rule. Your basic beliefs are on the bottom level B_0. Beliefs on this level are those that are not justified by other beliefs. You simply accept them as true. Perhaps they are observations or mathematical axioms.

Successor Rule. For every n, the beliefs on level B_n are beliefs that are justified by a good argument whose premises are beliefs on lower levels. A good argument is either a valid deductive argument or a logically acceptable inductive argument.

Final Rule. The collection that includes all beliefs on all levels. This collection is B. All your justified beliefs are in B. The collection B is the union of the B_n for all n.

Another philosophical example involves the objects to which we have epistemic access. Boyd says these objects are stratified into levels:

> Let O_1 be the class of entities that are observable to the typical unaided senses; for any n, let O_{n+1} be the class of entities that are detectable by procedures whose legitimacy . . . can be established without presupposing the existence of entities not in O_n; the union of the sets O_n is the class of observables in the sense relevant to the epistemology of science. (1984: 47)

The idea, roughly, is that the objects we can perceive with our unaided senses are on the bottom level. For example, you can perceive a piece of rounded glass with your naked eye. Objects we can perceive by means of objects on lower levels go on higher levels. For instance, using the piece of rounded glass as a lens, you can perceive microbes that you couldn't perceive before. That is, using objects on lower levels, you can make scientific instruments to detect objects on higher levels. We can express this by three rules:

Initial Rule. O_1 is the set of all objects perceivable by the unaided senses.

Successor Rule. O_{n+1} is the set of all objects detectable by means of scientific instruments built using objects on lower levels.

Final Rule. $O = \cup\{ O_n \mid n \text{ is a natural number } \}$.

2. Limits of Recursively Defined Series

2.1 Counting Through All the Numbers

It's easy enough to count through a few small numbers. You start counting by saying the numbers in order: 0, 1, 2, 3, 4, and so on. If you go on for a few minutes, you might count into the hundreds. But no matter how high you count, and no matter how long you take, there is still a bigger number you haven't counted to yet. No matter how long you go on counting, you can't count through all the numbers. That's obvious, right? No, it isn't obvious at all. What's obvious is that if you take the same amount of time to count off each next number, then it takes forever to count through all the numbers. But what if you don't take the same amount of time to count off each next number? What if you count off each number twice as fast as the one before? What happens then?

It's fun to *accelerate*. Here's how it works: It doesn't take you any time to count to 0. By default, you've counted to 0 when you start at time 0. Then you take 1/2 of a minute to say 1. Since you count twice as fast, you take 1/4 of a minute to say 2. Since you count twice as fast, you take 1/8 of a minute to say 3. Table 7.1 illustrates how long it takes you to accelerate through the numbers. Since you're always doubling your speed, your speed is increasing by powers of 2. Table 7.2 shows the pattern of your counting in terms of the powers of 2. For any number greater than 0, the time it takes you to say that number is $1/2^n$. And you've said that number by time $(2^n - 1)/2^n$.

Sadly, you can't really keep doubling your counting speed. Too bad. But what if you could? Suppose that you can accelerate. Suppose that for any number n, if it takes you some period of time to say n, then you can say the next number $n+1$ twice as fast. You start counting. As expected, you count to 1 by 1/2 of a minute; you count to 2 by 3/4 of a minute; you count to 3 by 7/8 of a minute. But what happens at 1 minute? For any number n, there is some time $(2^n - 1) / 2^n$ at which you've said it. Hence for any number n, there is some time less than one minute at which you've said it. By 1 minute, there are no numbers left for you to say. You have counted through all the numbers.

But maybe we've been imprecise. For the sake of precision, we should say that by 1 minute, there are no *finite* numbers left for you to say. So maybe there is a number for you to say at 1 minute. At 1 minute, you say *infinity*. By accelerating, by always going twice as fast, you've counted to infinity in exactly one minute. If you can accelerate, then infinity isn't something that you can't get to. It doesn't take you forever to count to it. On the contrary, if you can accelerate, then it only takes you a minute to count to infinity. You can get there, and you can get there quickly. Of course, you still might object that you can't actually accelerate. Fine. But if it is *possible* for you to accelerate, then it is *possible* for you to count to infinity in 1 minute. And why wouldn't it be possible? But suppose you insist that it is not possible for you. Still fine. For even if it is not possible *for you*, it is not logically impossible. It is possible for some agent to accelerate. Hence it is possible for some agent to count through all the finite numbers to infinity in 1 minute. In this sense, it can be done.

Number n	**Time in minutes that you take to count to n**	**Total time elapsed in minutes**
0	0	0
1	1/2	1/2
2	1/4	3/4
3	1/8	7/8
4	1/16	15/16
5	1/32	31/32
6	1/64	63/64

Table 7.1 Counting faster and faster.

Number n	**2 raised to the n-th power = 2^n**	**$1 / 2^n$**	**$2^n - 1$**	**$(2^n - 1) / 2^n$**
0	1	1/1	0	0
1	2	1/2	1	1/2
2	4	1/4	3	3/4
3	8	1/8	7	7/8
4	16	1/16	15	15/16
5	32	1/32	31	31/32
6	64	1/64	63	63/64

Table 7.2 Measuring your progress.

2.2 Cantor's Three Number Generating Rules

We don't really need to talk about agents who count faster and faster. While such talk is fun, it isn't essential. Modern thinkers have worked out a theory of infinity that's both extremely powerful and surprisingly easy to understand. The modern theory of the infinite begins with the Russian-German mathematician Georg Cantor in the late 1800s. Cantor used three rules to define a series of numbers that rises into the infinite (Hallett, 1988: 49). The three rules are: the initial rule; the successor rule; and the limit rule. They look like this:

> *Initial Rule*. There exists an initial number 0.
>
> *Successor Rule*. For every number n, there exists a successor number $n+1$. The successor rule generates all the positive finite numbers 1, 2, 3, and so on.
>
> *Limit Rule*. For any endless series of increasingly large numbers, there exists a limit number greater than every number in that series.

Since the initial and successor rules define an endless series of increasingly great numbers (the series 0, 1, 2, 3 and so on), there exists a limit number greater than every number in that series. Cantor gave the name ω to this first limit number. The symbol ω is the last letter of the Greek alphabet. It is pronounced "little omega". Every finite number is in the series that starts with 0 and that includes all the successors of 0. Since ω is greater than every number in that series, ω is greater than every finite number. And since ω is greater than every finite number, ω is infinite. More precisely, ω is the first *transfinite* number. But it's not the last transfinite number. The term "limit" does not imply the end. It implies a new beginning. Since ω is a number, it has a successor $\omega+1$. There is an endless series of transfinite numbers greater than ω. We'll discuss greater transfinite numbers soon. For now, all that's needed is the concept of the limit of a series.

2.3 The Series of Von Neumann Numbers

In the 20th century, the mathematician John von Neumann gave a nice recursive definition of numbers. He said each number n is the set of all numbers less than n. We can use von Neumann's definition to make Cantor's three rules precise:

> *Initial Rule*. There exists an initial number 0. Since n is the set of all numbers less than n, 0 is the set of all numbers less than 0. We're only talking about the natural numbers here, so negative numbers don't count. There are no natural numbers less than 0. Hence the set of such numbers is empty. Thus $0 = \{\}$.

Successor Rule. For every number n, there exists a successor number $n+1$. Since every number is the set of all lesser numbers, $n+1 = \{0, \ldots n\}$. Thus $1 = \{0\}$; $2 = \{0, 1\}$; $3 = \{0, 1, 2\}$; $4 = \{0, 1, 2, 3\}$; and so it goes.

Limit Rule. For the endless series of increasingly large finite numbers, there exists a limit number ω greater than every number in that series. The limit number ω is the set of all numbers less than ω. Every number defined by the initial and successor rules is less than ω. Hence every finite number is less than ω. So ω is the set of all finite numbers. It follows that $\omega = \{0, 1, 2, 3, \ldots\}$. As we said, there are numbers greater than ω. We'll deal with them in Chapter 8.

These rules generate numbers in a linear order – they generate a linear sequence of numbers, a number line. Clearly, the natural numbers are on this line. All natural numbers are finite. But the limit rule generates a number that is not a finite number – it is not a natural number. What kind of number is it? Since all these numbers are generated in linear order, we'll refer to all the numbers generated by these rules as *ordinal numbers*. The natural numbers are just the finite ordinal numbers. But ω is an infinite ordinal number. We'll talk more about ordinals in Chapter 8, section 3.

3. Some Examples of Series with Limits

3.1 Achilles Runs on Zeno's Racetrack

Zeno tells a story of Achilles running on a racetrack. Zeno says: Achilles is going to run a race on a straight flat racetrack. The racetrack is 1 mile long. The starting point is marked 0 miles, and the finish line is marked 1 mile. Achilles starts at time 0 at the starting point 0. We can picture Achilles as jumping from point to point along the racetrack. He takes a first jump that goes 1/2 the distance to the finish line in 1/2 minute. He takes a second jump that goes 1/2 of the remaining distance to the finish line in the next 1/4 minute. The total elapsed time is now 1/2 + 1/4 = 3/4 minutes. Likewise, the total distance covered is 3/4 of the way to the finish. He takes a third step that goes 1/2 the remaining distance to the finish line in the next 1/8 minute. The total elapsed time is 3/4 + 1/8 = 7/8 minutes. Likewise, again, the total distance covered is 7/8 of the way to the finish. He goes on according to this rule: he always takes a jump that is half the size of his last jump in half the time that it took to take his last jump. Thus Achilles accelerates.

The rules imply that at each time *less than* 1 minute, Achilles has not yet reached the finish. But where is Achilles at 1 minute after starting? When Zeno first described the movement of Achilles, he thought it was impossible for Achilles to get to the finish line by always jumping half way. He argued that wherever Achilles may be at a time less than 1 minute, he still has half way to go to get to the finish. Therefore he never arrives at the finish. As far as it goes, this story is fine. But it fails to go far enough – it fails to go the whole way. While it is true that Achilles is short of the finish at any time *less than* 1 minute, Zeno's reasoning says nothing about where he is at *exactly* 1 minute. Indeed, at 1 minute, the rule implies that Achilles *must be* at the finish. For the rule implies that at exactly 1 minute, Achilles has gone past every point less than 1 mile. Consequently: at 1 minute, Achilles is at the finish line – he is at distance 1. And, at 1 minute, he has taken as many jumps as there are natural numbers. It is the limit of his sequence of jumps.

Zeno Point. We'll use Z_n to indicate the position of Achilles on the racetrack after his n-th jump. The point Z_n is the n-th *Zeno point*. We can define the positions occupied by Achilles during the race – the Zeno points – using three rules:

Initial Rule. The initial Zeno point $Z_0 = 0$.

Successor Rule. For any n, the successor Zeno point $Z_{n+1} = (2^n - 1) / 2^n$.

Limit Rule. The limit Zeno point is 1. Since Achilles has already taken as many jumps as there are natural numbers, this limit Zeno point is Z_ω. Thus $Z_\omega = 1$.

Zeno Instant. Since the time it takes for Achilles to jump is the same as the distance he jumps (e.g., 1/2 mile in 1/2 minute), each Zeno point is equivalent to a *Zeno instant*. Just as we can divide a unit interval in space into infinitely many Zeno points, so we can divide a unit interval in time (e.g., 1 minute) into infinitely many Zeno instants.

3.2 The Royce Map

The 19th century American philosopher Josiah Royce describes an infinitely complex physical structure: a perfectly accurate map of England, located somewhere in England. It is clear that the definition is recursive: the perfectly accurate map is a thing in England that repeats the structure of England. Royce says:

> Whatever our theory of the meaning of the verb *to be*, suppose that some one . . . assured us of this as a truth about existence, viz., "Upon and within the surface of England *there exists* somehow (no matter how or when made) an absolutely perfect map of the whole of England." Suppose that . . . we had accepted this assertion as true. Suppose that we then attempted to discover the meaning implied in this one assertion. We should at once observe that in this one assertion, "A part of England perfectly maps all England, on a smaller scale," there would be implied the assertion not now of a process of trying to draw maps,

> but of the contemporaneous presence, in England, of an infinite number of maps, of the type just described. The whole infinite series, possessing no last member, would be asserted as a fact of existence. . . . the perfect map of England, drawn within the limits of England, and upon a part of its surface, would, if really expressed, involve, in its necessary structure, the series of maps within maps such that no one of the maps was the last in the series. (Royce, 1927: 506-507)

For simplicity, say England is just a square crossed by a north-south road and an east-west road. (England ain't what it used to be.) Figure 7.1 shows the first four iterations of the Royce map. We can describe it by these rules:

Initial Rule. The initial map M_0 = a square with a cross drawn in it.

Successor Rule. For any n, the successor map M_{n+1} = the map M_n + a cross drawn in the lower right square of M_n.

Limit Rule. The limit map M_ω = the super-imposition of all the M_n for n finite.

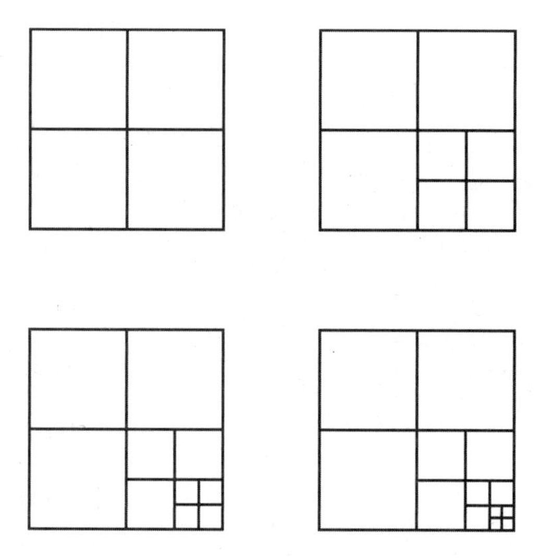

Figure 7.1 The first four iterations of a Roycean self-nested map.

3.3 The Hilbert Paper

A paper with all finitely long stroke series can be called the *Hilbert Paper*. The Hilbert Paper is infinitely complex: any square in the lower right hand corner has exactly the same structure as the whole Hilbert Paper. The first four iterations of the Hilbert Paper are shown in Figure 7.2. It is defined by these rules:

Initial Rule. The initial Hilbert Paper H_0 = a piece of paper divided in half vertically and horizontally with a single stroke | in the upper left quarter.

Successor Rule. For any n, the successor Hilbert Paper $H_{n+1} = H_n$ + you divide the right column in half vertically and you divide the bottom row in half horizontally; you copy the last row of strokes into the next lower row; you add one stroke on the right.

Limit Rule. The limit Hilbert Paper, which is the full Hilbert Paper, is H_ω = the super-imposition of all the H_n with n finite.

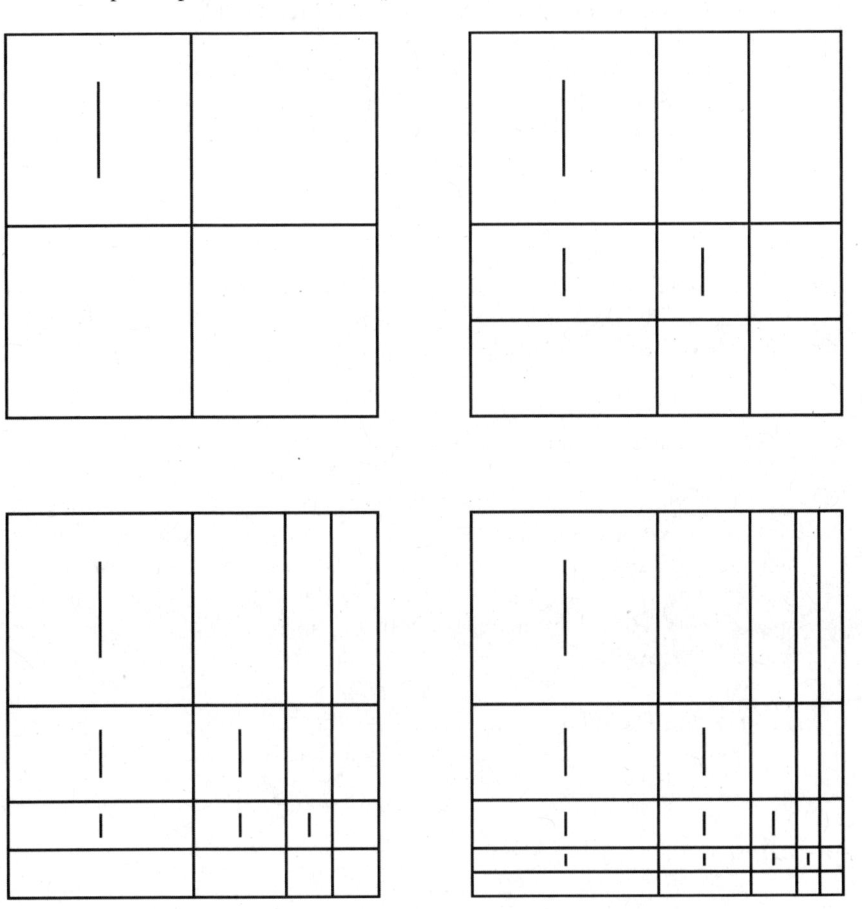

Figure 7.2 A few iterations towards the Hilbert Paper.

3.4 An Endless Series of Degrees of Perfection

You'll remember that Anselm argued for a finite series of degrees of perfection. After Anselm, the great chain of being became endless. As early as Locke, the hierarchy of increasingly perfect natures was thought to rise to the infinite. And the divine nature was thought to be greater than every nature in the endlessly rising hierarchy. So the perfection of the divine nature was truly infinite. Locke writes that

> in all the visible corporeal World, we see no Chasms or Gaps. All quite down from us, the descent is by easy steps, and a continued series of Things, that in each remove, differ very little one from the other. . . . And when we consider the infinite Power and Wisdom of the Maker, we have reason to think, that it is suitable to . . . the great Design and infinite Goodness of the Architect, that the Species of Creatures should also, by gentle degrees, ascend upward from us toward his infinite Perfection, as we see they gradually descend from us downwards: Which if it be probable, we have reason then to be persuaded, that there are far more Species of Creatures above us, than there are beneath; we being in degrees of Perfection much more remote from the infinite Being of GOD, than we are from the lowest state of Being. (Locke, 1690: III.6.12)

Of course, Locke didn't know about Cantor's notion of limits. He'd have to wait another two hundred years for that. But we can use the Cantorian notion to formalize the endless series of links in the great chain:

> *Initial Rule.* The initial degree of perfection is D_0. Let's stick with tradition and say it contains merely existing things. It's full of rocks.
>
> *Successor Rule.* For any n, the successor degree is D_{n+1}. As we rise through the continued series of things, we pass through degrees that contain plants, animals, humans, and who knows what else. Above humans we have – well, maybe angels, maybe super-intelligent creatures from other planets. It matters not. What does matter is that for every n, there exists a non-empty successor degree D_{n+1}.
>
> *Limit Rule.* The limit degree is D_ω. This degree is infinitely far above all the finite degrees. It contains an infinitely perfect Being, namely, GOD.

4. Infinity

4.1 Infinity and Infinite Complexity

A recursive definition is a finite way to describe a set whose cardinality (whose size) is greater than any finite number. Consider again the definition of the set of natural numbers N. It looks like this: 0 is in N; for every x in N, x+1 is in N; no other objects are in N. The definition uses only finitely many symbols. But N is larger than any finite set. Obviously, N is an infinite set of numbers. But not all infinite sets are sets of numbers. We need a clear and entirely general way to define infinite sets.

Infinite. A set S is *infinite* iff there exists a proper subset T of S such that the cardinality of T equals the cardinality of S. Equivalently, there is a bijection (a 1-1 correspondence) from S onto T. A set that satisfies this definition is also known as *Dedekind infinite.*

For example, consider the set of natural numbers N and the set of even numbers E. Since every even number is a natural number, E is a subset of N. And since there are natural numbers that are not even, E is a *proper* subset of N. The bijection f from N onto E is simple: since doubling each number gives an even number, just let $f(n) = 2n$. The bijection partly looks like this:

0	1	2	3	4	5	.	.	.	n
0	2	4	6	8	10	.	.	.	$2n$.

Doubling associates each number in N with an even number in E. And the inverse associates every even number in E with a number in N. So doubling is a bijection. Hence there are exactly as many even numbers as numbers. Analogously, there are as many odd numbers as numbers. A similar strategy shows that the even numbers are infinite too:

0	2	4	6	8	10	.	.	.	$2n$
0	4	8	12	16	20	.	.	.	$4n$.

And we can continue the doubling map endlessly. Hence the natural numbers is a self-representative system like Royce's map. It contains a copy of itself inside itself; the copy contains a copy; the copy contains . . . and so it goes.

Finite. A set S is *finite* iff it is not infinite. It is finite iff it does not contain any proper subset with the same cardinality.

Infinite Complexity. A structure S is *infinitely complex* iff S contains a proper part with exactly the same structure as itself. More formally, S is infinitely complex iff there exists a proper part T of S such that T has exactly the same form as S. The proper part T is a *proper substructure* of S. The part T has the same form as the whole S iff there exists an isomorphism from S to T. Equal form implies equal complexity: x has exactly the same structure as y implies that x is as complex as y. An infinitely complex structure is sometimes said to be *infinitary*. For example, consider the structure (N, $<$) where N is the natural numbers and $<$ is the less than relation. This is an ordered set. If E is the even numbers, then (N, $<$) is isomorphic to (E, $<$). Mapping n onto $2n$ preserves the order. Hence (N, $<$) is infinitary. Likewise Royce's self-nested map and the Hilbert Paper are infinitely complex.

Finite Complexity. A whole is *finitely complex* iff it is not infinitely complex. A whole is finite iff every part of the whole is less complex than the whole. A finitely complex whole or structure is sometimes said to be *finitary*.

4.2 The Hilbert Hotel

An infinitary structure has some strange properties. Ordinary common sense intuitions don't apply very well to infinitary structures. The failure of these intuitions can be nicely illustrated by an infinitary structure known as the *Hilbert Hotel*. The Hilbert Hotel consists of a very long hallway with numbered doors. There is a suite off the hall for each door. There is a suite for every natural number. Hence there are infinitely many suites. Let's suppose you're the Manager of the Hotel. You have to handle some tricky situations.

On Monday night the Hilbert Hotel is full. Every room is occupied by a guest, and the Hotel policy is that no suite can hold more than one guest. Late that same night, a new person shows up asking for a suite in the Hotel. As the Manager, what should you do? Should you turn her away? After all, every room contains a guest already. Although the Hotel appears to be full, the appearance is misleading. Here's the solution: you tell each guest to move down to the next suite. Thus the guest in suite 0 moves to suite 1; the guest in suite 1 moves to suite 2; and so on, so that for every suite number n, the guest in suite n moves into suite n+1. Now suite 0 is empty, and every guest who was in the Hotel still has a suite in the Hotel. You can easily put the newcomer into suite 0.

On Tuesday night the Hotel is still full. But now infinitely many guests arrive. They arrive on a single infinitary bus (how do they all fit?). There are as many new guests as natural numbers. Can you fit them all in? Of course you can. You just tell each guest already in the Hotel to double his or her suite number, and move down to that suite. For each guest, if he or she is in suite n, then he or she moves to suite $2n$. The set of even numbers has the same cardinality as the sets of numbers. So everybody who had a suite in the Hotel still has a suite in the Hotel. Now every odd numbered room is empty. Since there are as many odd numbers as numbers, you can put the newcomers into their suites.

On Wednesday night, the Hotel is full yet again. These people never leave. Now an infinite number of infinitary buses arrive. Each holds as many new arrivals as there are natural numbers. And there are as many infinitary buses as natural numbers. An infinity of infinities. Surely you can't squeeze them all in! But you can. Consider this: there are infinitely many prime numbers. A number is prime iff it is divisible only by itself and 1. A partial list of primes looks like this: 2, 3, 5, 7, 11, 13, 17. Now, for any prime number p, and any positive number n, consider p raised to the n-th power. Denote this p^n. The powers of 2 start like this: 2, 4, 8, 16, 32, 64, 128. The powers of 3 start like this: 3, 9, 27, 81, 243. The powers of 5 start like this: 5, 25, 125, 625. There are infinitely many primes. And for each prime, there are infinitely many powers of that prime. If two primes are distinct, their powers are distinct. Here we have our infinity of infinities. Start with the guests already in the Hotel. The guest in suite n moves to suite 2^n. The arrivals on the first bus move into the suites numbered by the powers of 3. Those on the second bus move into the suites numbered by the powers of 5. And for any m, those on the m-th bus move into the suites numbered by the powers of the m-th prime after 2.

4.3 Operations on Infinite Sequences

An *infinite sequence* is a function S from the set of natural numbers N to some set of objects T. As before, if S is an infinite sequence from N to the set of objects T, then S(n) is the n-th item in the sequence (it is the n-th item in T as ordered by S). We use a special notation and write S(n) as S_n. We write the sequence as $\{S_0, \ldots\}$.

Given an infinite sequence $\{S_0, \ldots\}$ of numbers, we can define the sum of its members by adding them in sequential order. The infinite sum looks like this:

$$\text{the sum, for } i = 0 \text{ to infinity, of } S_i = \sum_{i=0}^{\omega} S_i .$$

Given a sequence $\{S_0, \ldots\}$ of sets, we can define the union of its members by taking their union in sequential order. The infinite union looks like this:

$$\text{the union, for } i = 0 \text{ to infinity, of } S_i = \bigcup_{i=0}^{\omega} S_i .$$

Analogous remarks hold for infinite intersections:

$$\text{the intersection, for } i = 0 \text{ to infinity, of } S_i = \bigcap_{i=0}^{\omega} S_i .$$

5. Supertasks

5.1 Reading the Borges Book

Supertasks. A *supertask* is an infinite series of operations performed in a finite period of time and possibly a finite volume of space. Koetsier & Allis (1997) provide an excellent study of supertasks and give many examples. We've already considered a few supertasks. Assuming that all the guests in the Hilbert Hotel move simultaneously, or that their motions accelerate, then all the ways of moving the guests in the Hilbert Hotel were supertasks. And if you think of the infinite sums and unions as being done step by step, that is, if you think of them as computations, then they are supertasks. They are *super-computations*.

The *Borges Book* was written by the famous South American author Jorge Luis Borges. The Borges Book is not like ordinary books – it is infinitary. Its covers are thick. The front cover is 1/2 inch thick and the back cover is 1/2 inch thick. As the pages go inwards from the covers, they decrease in thickness by half. That is, they become twice as thin. The first page is 1/4 inches thick. Page 2 is 1/8 inches thick. And so it goes to the center of the book. The same rule applies from the back: the last page is 1/4 inches thick; the page before that is only 1/8 inches thick. And so it goes to the center of the book. The whole book is exactly 2 inches thick. How do you read the Borges Book?

Obviously, reading the Borges Book is a supertask. Suppose you try to read the Borges Book by accelerating from the front cover. You read the cover in 1/2 second; you read the first page in 1/4 second; and so on towards the middle. In one second, you've read the first half of the book. But the remaining pages remain unread. Of course, you could just repeat your acceleration from the back cover. That would work. But there's a more interesting way. You *oscillate* or alternate between front and back pages. You start with the front cover; then you read the back cover; then you read the first page; then you read the last page; then you read the second page; then you read the next to last page; and so on. As you read, you accelerate. You can read the whole book in one second.

5.2 The Thomson Lamp

The Thomson Lamp is a curious device (Thomson, 1954). It looks like an ordinary lamp with a single switch controlling a single light bulb. The difference is that you can switch it on or off at any speed. Start with the lamp off. You accelerate. In the first 1/2 second, you switch it on; in the next 1/4 second, you switch it off. You proceed to switch the lamp at the Zeno instants. We can describe the switching procedure by two rules:

> *Initial Rule.* At the initial Zeno instant Z_0, the lamp is off.
>
> *Successor Rule.* For any n, at the successor Zeno instant Z_{n+1}, the state of the lamp is the opposite of what it was at the previous Zeno instant Z_n.

We thus obtain an alternating sequence off, on, off, on, et cetera. The Thomson Lamp was originally presented as a paradox. The problem is this: what is the state of the lamp at 1 second? Is it on or off? The paradox is merely apparent. As Benacerraf (1962) observed, the procedure for switching the lamp on and off does not define any state of the lamp at 1 second. The procedure consists of an initial rule and a successor rule but no limit rule. *Any* limit rule is consistent with the initial and successor rules. Hence the lamp could be either on or off at time 1. You can define its state then any way you want.

The purpose of the Thomson Lamp example is to show that defining an initial rule and a successor rule does not entail anything about the limit of an infinite series of operations. To make sure, you need to explicitly define a limit rule.

5.3 Zeus Performs a Super-Computation

Zeus loves to compute. He has a finitely long tape divided into squares at Zeno points. The first square runs from 0 to 1/2; the second from 1/2 to 3/4; the third from 3/4 to 7/8; and so on to infinity. The tape is defined like this: there is a leftmost square of finite width; if x is any square of the tape, then there is a square of half the width of x to the right of x. This *Zeno tape* is exactly like the racetrack on which Achilles ran.

Zeus associates the leftmost square with the number 0; if Zeus associates some square with the number n, then he associates the next square to the right with the number $n+1$. The total number of squares on Zeus's tape is ω. The ω-th square on the tape is the limit square. It is a strange square. It has 0 width. It is as wide as a point. Zeus computes by writing numbers in squares. He uses his magic pencil. The tip of this pencil is exactly the size of a point. So if any square has a finite width, Zeus can write any finite sequence of digits in that square. Zeus can thus inscribe any finitely wide square with any finite number. He cannot write anything in the ω-th square. He can only rest his pencil point there. Since Zeus associates squares with numbers, when he computes he associates every natural number (the number of the square itself) with a natural number (the number written in that square). He is defining a function from the set N onto N.

Zeus uses his tape to determine the locations of the primes in the natural numbers. For any n, and for any finite interval of time, he is able to determine whether n is prime or not. He starts with a blank tape and his pencil at the leftmost square 0. He computes like this: for each square n, he determines whether n is prime or not. If it is prime, he writes 1 in that square; if it is not prime, he writes 0 in that square. He then moves right one square to determine whether or not $n+1$ is prime. He thus works through all the natural numbers. Of course, he accelerates. He determines whether n is prime by the n-th Zeno instant. When 1 unit of time has elapsed, his pencil is resting on the ω-th square at the rightmost end of the tape. Every square to the left of the ω-th square is marked with either 0 or 1. Zeus now has a nice list of the prime numbers to use in further computations. When Zeus computes the primes in this way, he is performing a supertask. It is a super-computation.

5.4 Accelerating Turing Machines

The image of Zeus computing with his tape should remind you of the Turing machines from Chapter 3. A Turing machine (TM) is a read-write head that moves back and forth over an endless tape. TMs have some well-known limitations. Consider a famous problem known as the *Halting Problem*. This problem is based on an interesting feature of TMs. The behavior of any TM can be encoded in a number – its program. And the input to any TM is also a number. You can thus imagine a table, call it the *Halting Table*, whose columns are numbered with programs and whose rows are numbered with inputs. The cell in column n and row m is 1 if the TM with program number n halts when given input number m. That cell contains a 0 if it does not halt. Turing proved that no TM can solve the Halting Problem. No single TM can fill in the Halting Table with 0s and 1s in the right way. And TMs have other limitations (see Boolos & Jeffrey, 1989: chs. 4 & 5).

There are various ways the limitations of TMs can be overcome. One way is to let them accelerate. Any TM performs a single operational cycle in one time step: it reads the tape; it changes its state; it moves left or right or changes the tape. There's nothing in the abstract definition of a TM that requires it to spend the same amount of time on each operational cycle. It can do its first operational cycle in 1/2 second; its next in 1/4 second; and so on. It can thus complete an infinite series of operational cycles in 1 second, much like Zeus working out his tape with all the primes marked with 1s. An *accelerating Turing machine* (an ATM) can accelerate. It can perform supertasks. We won't go into the details of ATMs here. Copeland (1998a) is an excellent introduction. Remarkably, Copeland (1998b) shows that an ATM can solve the Halting Problem. There is a precise sense in which ATMs are more powerful than TMs.

Some philosophers say that a function is *computable* iff it is computable by a TM – so that the Halting Problem is not computable. This notion of computability enters into debates about the powers of human minds. One argument goes like this: (1) minds can do calculations that cannot be done by TMs; (2) but anything that can be done by a computer can be done by a TM; therefore (3) minds are more powerful than computers.

There's a purely formal problem with this argument – it really isn't accurate to identify computability with what can be computed by a TM. ATMs are computers, and they're more powerful than TMs. One kind of computability is whatever a TM can do – we can call it *Turing computability*. Another kind of computability is whatever an ATM can do. Given the parallels of an ATM with Zeus, we might call it *Zeus computability*. So even if minds can do more than TMs, they still might be computers. We won't go into these issues any further here. We just want to show you how supertasks have some philosophical relevance. You can find an excellent discussion in Copeland (2000).

8

BIGGER INFINITIES

1. Some Transfinite Ordinal Numbers

When we talk about ordinal numbers, we always use the von Neumann definition of numbers. Thus any ordinal number n is the set of all numbers less than n. This works for finite numbers. For example, $5 = \{0, 1, 2, 3, 4\}$. And it works for infinite numbers. The infinite number ω is the set of all finite numbers: $\omega = \{0, 1, 2, 3, \ldots\}$.

Since ω is infinite, it might seem like ω is the end of the number line. It might seem like ω has no successors. But Cantor's three number generating rules are entirely general. So it follows that ω has a successor. After all, ω is a number, and the successor rule says that *every* number n has a successor n+1. So the successor of ω is ω+1. What could this number be? We apply the von Neumann definition:

n = the set of all numbers less than n;

ω+1 = the set of all numbers less than ω+1.

We already know that every finite number is less than ω. Hence every finite number is less than ω+1. And the successor rule tells us that ω is less than ω+1. It follows that ω+1 is the set of all x such that either x is a finite number or x is equal to ω. In symbols,

$\omega+1 = \{ x \mid x \text{ is finite or } x = \omega \}$;

$\omega+1 = \{0, 1, 2, \ldots \omega\}$.

By the same reasoning, we can define a successor of ω+1. That is:

$\omega+2 = \{0, 1, 2, \ldots \omega, \omega+1\}$.

As expected, we can repeat this process endlessly:

$\omega+3 = \{0, 1, 2, \ldots \omega, \omega+1, \omega+2\}$;

$\omega+4 = \{0, 1, 2, \ldots \omega, \omega+1, \omega+2, \omega+3\}$.

This is an endless series defined by addition:

$\omega, \omega+1, \omega+2, \omega+3, \omega+4, \ldots \omega+n, \ldots.$

Since we now have an endless progression of ever-greater numbers after ω, we can apply Cantor's limit rule to get another limit number:

$\omega+\omega = \{0, 1, 2, \ldots \omega, \omega+1, \omega+2, \ldots\}$.

By analogy with the finite numbers, we can say $\omega+\omega = \omega \cdot 2$. The same reasoning that led us from ω to $\omega \cdot 2$ leads us from $\omega \cdot 2$ to $\omega \cdot 3$. The result is an endless series defined by multiplication:

$\omega, \omega \cdot 2, \omega \cdot 3, \omega \cdot 4, \ldots \omega \cdot n, \ldots$

The limit of the multiplicative series above is $\omega \cdot \omega$. Again, by analogy with the finite numbers, we can say that $\omega \cdot \omega$ is ω^2. We now have an endless series defined by exponentiation:

$\omega, \omega^2, \omega^3, \omega^4, \ldots \omega^n, \ldots$

And the limit of this series is ω^ω. As you might expect, we can apply the Cantorian rules to generate even greater numbers. But there's little point in writing them down. We've shown, at least in an informal sense using Cantor's rules, that the line of numbers does not end with ω. It keeps going, and going, and going, and going . . .

2. Comparing the Sizes of Sets

It is possible to compare the sizes of sets without counting. It is often easier to perform the comparison without counting than to do it by counting. Consider a classroom filled with some desks and with some students. You can tell without counting whether there are (1) as many students as desks; (2) more students than desks; or (3) more desks than students. There are more desks than students if and only if every student is seated at one desk but not every desk is occupied by some student. There are more students than desks if and only if every desk is occupied by one student but not every student is seated at some desk. There are as many students as there are desks if and only if every student is seated at one desk and every desk is occupied by one student.

Less Than or Equal Cardinality. The *cardinality* of a set is its size. The symbolism $X \precsim Y$ means that the size of set X is less than or equal to the size of set Y. We define the relation $\precsim$ like this: $X \precsim Y$ iff there is some way to pair off every member of X with exactly one member of Y. If X $\precsim$ Y, then every member of X is paired off with exactly one member of Y, but there might be some members of Y that are not paired off with members of X. Every member of X has a unique partner in Y; but it is possible that there are members of Y without partners in X. We can define $\precsim$ in terms of functions: X $\precsim$ Y iff there is a 1-1 function f from X *into* Y. f is a 1-1 function from X *onto* a *subset* of Y.

Equal Cardinality. Suppose X $\precsim$ Y and Y $\precsim$ X. If that's true, then there is a way to pair off each member of X with exactly one member of Y and there is a way to pair off each member of Y with exactly one member of X. Consequently: every member of X has a partner in Y and every member of Y has a partner in X. There are as many things in X as there are things in Y. There is a 1-1 correspondence between X and Y. The symbolism X $\approx$ Y means that X is the same size as Y. We say X $\approx$ Y iff X $\precsim$ Y and Y $\precsim$ X. When X is the same size as Y, we also say X *is equinumerous with* Y or X *is equicardinal with* Y.

For example, the function $f(n) = 2n$ is a 1-1 correspondence between the set of numbers and the set of even numbers. It pairs each number n with an even number $2n$ and, conversely, each even number $2n$ with the number n. This shows that the set of even numbers is the same size as the set of numbers.

There is a 1-1 correspondence between ω and ω+1. We just pair off each positive n in ω with n-1 in ω+1, and we pair off 0 in ω with ω in ω+1. It looks like this:

1	2	3	.	.	.	n+1	.	.	.	0
0	1	2	.	.	.	n	.	.	.	ω.

This shows that the size of ω+1 is equal to the size of ω. By analogous informal reasoning, there is a 1-1 correspondence between ω and ω·2. Here it is:

0	2	4	6	.	.	.	1	3	5	7	.	.	.
0	1	2	3	.	.	.	ω	ω+1	ω+2	ω+3	.	.	.

And this shows informally that the size of ω·2 is equal to the size of ω. Although we will not prove it here, any number derived from ω by any series of arithmetical operations has the same size as ω itself.

We are now able to define all size comparisons between sets:

X is the same size as Y	iff X $\precsim$ Y and Y $\precsim$ X;
X is smaller or the same size as Y	iff X $\precsim$ Y;

X is smaller than Y	iff $X \precsim Y$ but not $Y \precsim X$;
X is larger or the same size as Y	iff $Y \precsim X$;
X is larger than Y	iff $Y \precsim X$ but not $X \precsim Y$.

3. Ordinal and Cardinal Numbers

Ordinary English distinguishes between *ordinal* and *cardinal* numbers. Ordinal numbers indicate order. The sequence of ordinal number words is *first*, *second*, *third*, *fourth*, and so on. Cardinal numbers indicate size (that is, amount or quantity). The sequence of cardinal number words is *one*, *two*, *three*, *four*, and so on. Ordinal and cardinal numbers are just two different ways of thinking of the natural numbers. More precisely, the *n*-th ordinal is identical with the *n*-th cardinal is identical with the number *n*. For example, *second* and *two* are both identical with 2. But we need to be more precise.

Ordinal Number. We've used Cantor's three number generating rules to informally define the ordinal numbers (the ordinals). Accordingly, 0 is an ordinal; for any *n*, if *n* is an ordinal, then the successor of *n* is an ordinal; and finally, if S is any endless series of increasing ordinals, then the limit of S is an ordinal.

Cardinal Number. Every set has some cardinal number. Its cardinal number is its cardinality. For any set S, the *cardinal number* of S is the smallest ordinal number that is the same size as S. It is the smallest ordinal that is equinumerous with S. Informally, we can say that an ordinal *n* is a cardinal iff the cardinal number of *n* is *n* itself.

According to our definition, every finite ordinal is a cardinal. For any finite ordinal *n*, the smallest ordinal that can be put in a 1-1 correspondence with *n* is *n* itself. The least infinite ordinal ω is also a cardinal. Since every ordinal *n* less than ω is finite, there cannot be any 1-1 correspondence between any finite *n* and ω. So the cardinality of ω is ω. The smallest ordinal with the same size as ω is ω itself. Thus ω is a cardinal number. Just as we can refer to 2 as an ordinal (by the word *second*) or to 2 as a cardinal (by the word *two*), so we can refer to ω as an ordinal or to ω as a cardinal. Here's how it's done:

ω is $\{0, 1, 2, \ldots\}$ thought of as an ordinal;

$\aleph_0$ is $\{0, 1, 2, \ldots\}$ thought of as a cardinal.

The symbol $\aleph$ is the first letter of the Hebrew alphabet. It is verbalized as *aleph*. So $\aleph_0$ is verbalized as *aleph-zero* or *aleph-naught*. Thus we say that the cardinality of ω is $\aleph_0$. The two different symbols ω and $\aleph_0$ are two different names for the same set. Hence $\omega = \aleph_0$. And both are just signs for the set of natural numbers N. So $\omega = \aleph_0 = N$.

For finite numbers, ordinality and cardinality are equivalent. Every cardinal is an ordinal and every ordinal is a cardinal. Different ordinals have different cardinalities. However, when we consider infinite ordinals, the concept of cardinality diverges from the concept of ordinality. Many different ordinals have the same cardinality.

For example, the cardinality of ω is $\aleph_0$. But the cardinality of $\omega+1$ is also $\aleph_0$. There is a 1-1 correspondence between ω and $\omega+1$. We just pair off each n in ω with $n+1$ in $\omega+1$, and we pair off 0 in ω with ω in $\omega+1$. Every ordinal that we can derive from ω by any arithmetic operations has the same cardinality as ω. It has cardinality $\aleph_0$. Table 8.1 shows some transfinite ordinals with cardinality $\aleph_0$.

Denumerable. We say a set S is *denumerable* iff it has the same cardinality as ω. That is, S is denumerable iff the cardinality of S is $\aleph_0$. Equivalently, S is denumerable iff there is a 1-1 correspondence between S and the set of natural numbers N.

Countable. We say a set S is *countable* iff either S is finite or S is denumerable. That is, a set is countable if it has finite cardinality or cardinality $\aleph_0$. The ordinal ω is the least countable infinite ordinal.

Let's define some 1-1 correspondences between ω and some greater countable ordinals. Obviously, ω can be put into a 1-1 correspondence with itself. And we know that ω can be put into a 1-1 correspondence with $\omega+1$ like this:

0	1	2	. . . ω
1	2	3	. . . 0.

We can put ω into a 1-1- correspondence with $\omega+2$ like this:

0	1	2	. . . ω	$\omega+1$
2	3	4	. . . 0	1.

And we can likewise generate a 1-1 correspondence between ω and $\omega+n$ for any finite n. By pairing off the n-th even number with $0+n$ and the n-th odd number with $\omega+n$, we get a 1-1 correspondence between ω and $\omega+\omega$. Here it is:

0	1	2	. . . ω	$\omega+1$	$\omega+2$	. . .
0	2	4	. . . 1	3	5	. . .

By working with multiples of 3 we can generate a 1-1 correspondence between ω and ω·3. It looks like this:

0, 1, 2 . . .	ω	ω+1	ω+2. . .	ω·2	ω·2+1	ω·2+2	. . .
0, 3, 6 . . .	1	4	7 . . .	2	5	8	. . .

By working in a similar way with multiples of n, we can generate 1-1 correspondences between ω and ω·n for any finite n. With a little effort, we can generate a 1-1 correspondence between ω and ω·ω. We have ω-many series, each with ω-many numbers in it. For the first series, use powers of the first prime 2. These are 2^1, 2^2, 2^3, 2^4 . . . For the second series, use powers of the second prime 3. These are 3^1, 3^2, 3^3, 3^4 . . . For the third series, use powers of the third prime 5. These are 5^1, 5^2, 5^3, 5^4 . . . For the n-th series, use powers of the n-th prime. There are ω-many primes; there are ω-many powers of each prime, so we have a map from the natural numbers ω to ω·ω.

Since the cardinality of ω is $\aleph_0$, each of these 1-1 correspondences shows that the greater ordinal also has cardinality $\aleph_0$. As a rule, mere arithmetical operations on ω will never make a number with cardinality greater than $\aleph_0$. This is shown in Table 8.1.

Ordinal	**Size**
ω	$\aleph_0$
ω+1	$\aleph_0$
ω+2	$\aleph_0$
ω+ω	$\aleph_0$
ω·3	$\aleph_0$
ω·ω	$\aleph_0$

Table 8.1 Some countably infinite sets.

4. Cantor's Diagonal Argument

The ordinal ω has size $\aleph_0$. Its cardinality is $\aleph_0$. And Table 8.1 shows several other ordinals, all greater than ω, but all with size $\aleph_0$. Table 8.1 concludes with the claim that arithmetical operations on ω will never make a number with cardinality greater than $\aleph_0$. Of course, all these numbers are just sets. So Table 8.1 suggests that all infinite sets have the same size – they all have size $\aleph_0$. But Table 8.1 doesn't offer any *proof* of that suggestion. In fact, the suggestion is false – there are sets whose sizes are greater than $\aleph_0$. This is another strange feature of infinity. It's natural to think that infinity is just infinity – all infinite sets must have the same size. But they don't. Some infinities are bigger than others.

A classical argument for the existence of infinite sets larger than $\aleph_0$ is *Cantor's Diagonal Argument.* Cantor's Diagonal Argument shows that some infinities are bigger than others. Start with the set of finite numbers ω. This set is also $\aleph_0$. How many ways are there to select numbers from ω? Every selection of numbers from ω is an infinitely long sequence of yes / no choices. We can display such selections in a table. The columns are labeled with the finite numbers. A selection includes the number *n* if the cell under column *n* is 1; it does not include *n* if the cell under *n* is 0. Table 8.2 partially illustrates a selection. The selection contains 2, 4, 5, and 7; it does not include 0, 1, 3, and 6. For every other finite number, the selection either includes or excludes it; but we don't have room to show that. If the selection were written on a Zeno tape, we could display it here.

n	0	1	2	3	4	5	6	7	. . .
Selection	0	0	1	0	1	1	0	1	. . .

Table 8.2 Part of a sample selection of finite numbers.

How many selections are there? Let's consider this more precisely. A selection is a function *f* from the set ω of finite numbers into the set {0, 1}. Each of these functions is a characteristic function over the set of finite numbers. So the set of selections is the set of characteristic functions $F = \{ f \mid f: \omega \rightarrow \{0, 1\}\}$. How big is F? More technically, what is the cardinality of F? Notice that F must be at least as big as $\aleph_0$. To see this, just pair each number *n* with the selection *f* in which $f(n)$ is 1 and every other value of *f* is 0. This pairing is a 1-1 function from $\aleph_0$ *into* F. So we know that the cardinality of F is either greater than or equal to the cardinality of $\aleph_0$. Suppose the cardinality of F is equal to $\aleph_0$. If that is true, then there is some 1-1 correspondence between F and $\aleph_0$. Since $\aleph_0 = \omega = \{0, 1, 2, \ldots\}$, there is some 1-1 correspondence between the set of finite numbers {0, 1, 2, . . .} and F. One of these correspondences might look like this:

0 ↔ 10110101010101. . .
1 ↔ 00001010000111. . .
2 ↔ 00000101001001. . .
3 ↔ 11110111111111. . .
4 ↔ 11111110000000. . .
and so on.

Given any alleged correspondence between $\aleph_0$ and F, we can put that entire correspondence into a table. Table 8.3 partially illustrates an alleged correspondence. The columns in Table 8.3 (after the first) are the numbers in $\aleph_0$. The rows in Table 8.3 (after the first) are the selections. A row-column cell is 1 if the selection in the row includes the number in the column; it is 0 if the selection in the row excludes the number in the column.

The Diagonal Argument shows that the set of selections of numbers is bigger than the set of finite numbers. It shows that the size of F is greater than $\aleph_0$. The size of F is a bigger infinity than $\aleph_0$. Let's spell out this argument in detail:

1. Assume: There is some 1-1 way to pair off selections in F with numbers in $\aleph_0$. The argument will proceed to derive a contradiction from this assumption.

2. If there is such a way, then *all* the selections can be put into a table like Table 8.3. Table 8.3 has exactly as many rows as there are numbers in $\aleph_0$.

3. Given any table of selections, make a *diagonal* selection by *negating* the n-th selection value from the n-th column in the n-th row, that is, change each 1 to a 0 and each 0 to a 1. For example, as we go down the diagonal in Table 8.3, we have the selection 10011. . . Negating these values, we get the diagonal selection 01100 . . .

4. The diagonal selection can't be in Table 8.3. It can't be in row 1 because its 1st value differs from the 1st value of row 1; it can't be in row 2 because its 2nd value differs from the 2nd value of row 2; . . . it can't be in row n because its n-th value differs from the n-th value of the n-th row. So the diagonal selection can't occur in any row in Table 8.3. And thus it can't be in Table 8.3 at all. Hence Table 8.3 does *not* contain all the selections. At least one selection, the diagonal selection, is missing from Table 8.3.

5. The reasoning is general. No matter how you try to pair numbers 1-1 with selections of numbers, you can always form a diagonal selection that doesn't occur in your pairing. So the assumption that you can pair selections 1-1 with numbers is wrong.

6. Conclusion: the set of selections is bigger than the set of numbers. The size of F is greater than the size of $\aleph_0$. The size of F is a bigger infinity.

	0	1	2	3	4	...
0	**1**	0	1	1	0	...
1	0	**0**	0	0	1	...
2	0	0	**0**	0	0	...
3	1	1	1	**1**	0	...
4	1	1	1	1	**1**	...
...	...	...	...	...	...	...

Table 8.3 The table of infinite selections.

The numbers in Table 8.3 have some interesting properties. Look at row 0. It contains selection 0. Now look at column 0 in row 0. It has a 1. This indicates that the number 0 is included in selection 0. Say a number is *happy* iff it is included in its own selection. That is, number *n* is happy iff row *n* and column *n* is 1. For example, in Table 8.3, the numbers 0, 3, and 4 are happy. Numbers that aren't happy are sad. They are matched up with selections that don't include them! That's sad. We say a number *n* is *sad* iff row *n* and column *n* is 0. The sad numbers in Table 8.3 are 1 and 2. Now suppose we form the selection of all sad numbers. The *sad selection* includes all the sad numbers and excludes all the happy numbers. The sad selection is just the selection 01100.... And that's the diagonal selection – the selection that isn't in Table 8.3.

We can use the distinction between happy and sad numbers to show – in a different way – that there is no 1-1 correspondence between numbers and selections of numbers. Consider the sad selection. If some table lists a 1-1 correspondence between ω and F, then it must contain the sad selection. For instance, suppose the sad selection 01100. . . is in Table 8.3. If the sad selection is in Table 8.3, then it appears in some row; hence there is some number *n* such that row *n* is the sad selection. And that row *n* has a *n*-th column. Either the cell in row *n* and column *n* is 0 or else it is 1. Call this the *crazy cell.* If the crazy cell is 0, then the number *n* is sad; but in that case, *n* must be in the sad selection, so that the crazy cell must be 1. If the crazy cell is 1, then *n* is happy; but if it is happy, then *n* cannot be in the sad selection (which does not contain any happy numbers); so the crazy cell must be 0. We can summarize: (1) if the sad selection is in some table, then that table contains a crazy cell; (2) if the crazy cell is 0, then it is 1; if it is 1, then it is 0. But (3) no table can contain such a cell. And therefore, (4) no table can contain the sad selection. Finally (5) if no table can contain the sad selection, there is no 1-1 correspondence between numbers and selections. The size of the set of selections is greater than the size of the set of numbers.

5. Cantor's Power Set Argument

5.1 Sketch of the Power Set Argument

The Diagonal Argument shows that there are more selections of numbers in $\aleph_0$ than numbers in $\aleph_0$. As we already know, a selection of numbers in $\aleph_0$ is a subset of $\aleph_0$. So the set of subsets of $\aleph_0$ is bigger than $\aleph_0$. For any set S, the set of all subsets of S is the power set of S. The power set of S is written *pow S*. Hence the set of all subsets of $\aleph_0$ is pow $\aleph_0$. The Diagonal Argument shows that for the single case of $\aleph_0$, the power set of $\aleph_0$ is bigger than $\aleph_0$. Can we generalize this reasoning? Can we prove that for *any* set S, the size of pow S is greater than the size of S? We can prove it. Cantor's *Power Set Argument* shows that for any set S, the size of pow S is greater than the size of S.

A little warm-up will be useful here. Since we've shown that pow ω is bigger than ω, let's talk a little about the subsets of ω. The set of all subsets of ω is pow ω. The *members* of pow ω are the *subsets* of ω. Pow ω contains all sets of natural numbers. For any natural number n, the set $\{n\}$ is in pow ω. This proves that pow ω is infinite. Every finite set of numbers is in pow ω. And every infinite set of numbers is in pow ω. The set of all even numbers is in pow ω. The set of all odd numbers is in pow ω. The set of all square numbers {1, 4, 9, 16, 25, 36, . . .} is in pow ω. Pow ω is a very large set.

If pow ω is the same size as ω, then there is a 1-1 correspondence f that associates every number n in ω with some unique set of numbers $f(n)$ in pow ω. Consider any 1-1 correspondence f that maps a number onto a set of numbers. Say a number n is *happy* iff n is in the set $f(n)$ and n is *sad* iff n is not in $f(n)$. Table 8.4 gives examples for the first seven numbers.

n	**Set of Numbers $f(n)$**	**Membership**	**Emotion**
0	{143098}	$0 \notin \{143098\}$	sad
1	{3}	$1 \notin \{3\}$	sad
2	{1, 2, 7}	$2 \in \{1, 2, 7\}$	happy
3	{3, 17, 24}	$3 \in \{3, 17, 24\}$	happy
4	{1, 4, 9, 16, 25, . . .}	$4 \in \{1, 4, 9, 16, 25, \ldots\}$	happy
5	{0, 2, 4, 6, 8, . . . }	$5 \notin \{0, 2, 4, 6, 8, \ldots\}$	sad
6	{14, 29}	$6 \notin \{14, 29\}$	sad

Table 8.4 Numbers that are "happy" or "sad".

For any pairing of numbers to sets of numbers, we can collect all the sad numbers. From Table 8.4 we get {0, 1, 5, 6, . . .}. Obviously, the set of sad numbers is a set of numbers (it may be finite or infinite). So it has to appear someplace in any 1-1 correspondence between numbers and sets of numbers. So there has to be some number Y that is matched by f with the set of all sad numbers. In other words, there is some Y such that f(Y) is the set of all sad numbers. And either Y is in f(Y) or Y is not in f(Y). If Y is in f(Y), then Y is happy. But if Y is not in f(Y), then Y is sad. Which is it? Table 8.5 asks this question.

n	**Set of Numbers *f*(*n*)**	**Emotion**
0	{143098}	sad
1	{3}	sad
2	{1, 2, 7}	happy
3	{3, 17, 24}	happy
4	{1, 4, 9, 16, 25, . . .}	happy
5	{0, 2, 4, 6, 8, . . . }	sad
6	{14, 29}	sad
. . .	. . .	. . .
Y	{0, 1, 5, 6, . . .}	*happy or sad?*
. . .	. . .	. . .

Table 8.5 The set of sad numbers.

Is our mystery number Y happy or sad? We have two cases: On the one hand, suppose Y is happy. If Y is happy, then Y is in its matched set f(Y); but it's matched set is the set of all *sad* numbers, so if Y is in that set, then Y is sad. If Y is happy, then Y is sad. On the other hand, suppose Y is sad. If Y is sad, then the set of all sad numbers f(Y) includes Y. And now, since Y is in f(Y), this means that Y is happy. If Y is sad, then Y is happy.

Our result is that Y is happy if and only if Y is sad. But that's absurd. So Y is neither happy nor sad. It follows that the 1-1 correspondence f cannot associate any number Y with the set of sad numbers. But for every 1-1 correspondence between ω and pow ω, there is a set of sad numbers. The lesson is this: no matter how you try to pair off numbers with sets of numbers, you can't pair off the set of sad numbers with any number. There is no 1-1 correspondence between ω and pow ω. We now know that the size of pow ω is not the same as the size of ω. And since we already know that pow ω is infinite, we know that the size of pow ω cannot be less than the size of ω. Only one option remains: the size of pow ω is greater than the size of ω. There are more sets of numbers than numbers.

5.2 The Power Set Argument in Detail

We illustrated Cantor's Power Set Argument using numbers. But the Argument is fully general – it works with any set. Here's the fully general version of Cantor's Power Set Argument in sequential format:

1. Consider *any* set A.

2. Assume that the size of A is equal to the size of the power set of A.

3. If A has as many members as the power set of A, then there is some 1-1 correspondence f that pairs each x in A with its own member of the power set of A and pairs every member of the power set of A with its own member of A.

4. For each x in A, $f(x)$ is a subset of A. That is, the members of $f(x)$ are also members of A. Hence for each x in A, either x is in $f(x)$ or not.

5. If x is in $f(x)$, then x is said to be *happy*. If x is not in $f(x)$, then x is said to be *sad*. Since any x in A is either in $f(x)$ or not, no x in A is both happy and sad.

6. Let F be the set of all x in A such that x is sad. Since every member of F is a sad member of A, every member of F is a member of A; so F is a subset of A; so F is a member of the power set of A.

7. Since f is supposed to put A and its power set into 1-1 correspondence, there is some y in A such that $f(y)$ is F. In other words: there is some y in A such that f maps y to the set of all sad members of A. Is y happy or sad?

8. On the one hand, suppose y is happy; if y is happy, then y is in $f(y)$; but since $f(y)$ is F, and since all the members of F are sad, it follows that y is sad.

9. On the other hand, suppose y is sad; if y is sad, then since $f(y)$ contains all sad sets, y is in $f(y)$; but if y is in $f(y)$, it follows that y is happy.

10. Therefore, y is sad if, and only if, y is happy. But no set is both happy and sad. We have arrived at a contradiction.

11. We must reject the assumption that f is a 1-1 correspondence between A and the power set of A. Since A and f could be any set and correspondence, there is never any 1-1 correspondence between any set A and the power set of A.

12. Consequently, the size of A is not equal to the size of the power set of A.

13. There is a 1-1 function from A *into* the power set of A. This function associates every x in A with its unit set $\{x\}$ in the power set of A. Therefore, the size of A is either less than or equal to the size of the power set of A.

14. Since the size of A is not equal to the size of its power set, the size of A must be strictly less than the size of its power set. Equivalently, the size of the power set of A is strictly greater than the size of A.

5.3 The Beth Numbers

We started with one infinite set: ω. We then defined two bigger infinite sets. The Diagonal Argument showed that the set of functions from ω to {0, 1} is bigger than $\aleph_0$. Formally, this set is $F = \{ f \mid f\colon \omega \rightarrow \{0, 1\}\}$. The Power Set Argument showed that the power set of ω is bigger than $\aleph_0$. Formally, pow $\omega = \{ x \mid x \subseteq \omega\}$. Every function in F corresponds to a single subset of ω and vice versa. So there is a 1-1 correspondence between F and pow ω. Consequently, F and pow ω have the same size. This size is bigger than $\aleph_0$. What is this size? Mathematicians use the symbol $\aleph$ to define this size. Be careful: $\aleph$ is just the aleph symbol, without any subscript. $\aleph$ is *not* identical with $\aleph_0$. $\aleph$ is the cardinality of pow ω and is also the cardinality of F. $\aleph$ is bigger than $\aleph_0$.

Uncountable. We say a set S is *uncountable* iff the cardinality of S is greater than $\aleph_0$. The set F of all selections over ω is uncountable. The set of all subsets of ω is uncountable. The cardinal number $\aleph$ is an uncountable transfinite number.

We can use the power set operation to define an endless series of bigger and bigger infinite sets. It's an axiom of set theory that for every set S, the power set of S exists and is also a set. So we can recursively define an endless series of increasingly large transfinite sets:

B(0) = ω;
B(n+1) = pow B(n).

For example,

B(0) = ω;
B(1) = pow ω;
B(2) = pow pow ω;
B(3) = pow pow pow ω;
and so it goes.

For finite numbers, the size of the power set of n is 2^n. We can write this on a single line of text using the notation 2^*n*. You read this as *2 raised to the n-th power*. We are free to extend this notation to infinite numbers. The size of the power set of any infinite number x is 2^*x*. So the size of power set of $\aleph_0$ is 2^$\aleph_0$. We can use this notation to express the sizes of the sets in the B series:

the size of B(0) = $\aleph_0$;

the size of B(n+1) = 2^the size of B(n).

We said that B(1) is pow ω. Hence the size of B(1) is the size of pow $\aleph_0$. We used $\aleph$ to denote this size. Thus the size of B(1) is $\aleph$. But we also said that the size of B(1) is 2^$\aleph_0$. It follows that $\aleph$ = 2^$\aleph_0$. Table 8.6 shows some of the B(n) sets and their sizes. The numbers formed as the sizes of the B(n) sets are the *beth numbers*. Every beth number is a transfinite cardinal. Every next beth number is bigger than the previous beth number. The series of beth numbers is an endlessly increasing series of ever bigger infinities.

You might worry that we've gone way too far out into the heaven of Platonic abstractions here. How could such esoteric objects as the beth numbers play any role in concrete reality? Well, the number $\aleph$ is closely connected to the idea of continuity. The number of points on a continuous line is $\aleph$. Equivalently, the size of the set of real numbers is $\aleph$. More physically, the number of points in any continuous space-time is $\aleph$. Physical theories can involve these big infinities. The set of all regions of some set of space-time points is the power set of that set of points. So if the number of space-time points is the size of B(1), then the number of space-time regions is the size of B(2). So at least some of these infinities play roles in physical theories. More philosophically, you might ask: how many possible universes are there with the same laws as our universe? That will be a pretty big infinity. Going further into possibility will generate even bigger infinities.

Set	**Size**
ω	$\aleph_0$
pow ω	2^$\aleph_0$
pow pow ω	2^(2^$\aleph_0$)
pow pow pow ω	2^(2^(2^$\aleph_0$))
pow pow pow pow ω	2^(2^(2^(2^$\aleph_0$)))

Table 8.6 Iterated power sets and their sizes.

6. The Aleph Numbers

Alephs. An *aleph* is an infinite cardinal number. Since an aleph is beyond the finite, we can also say it is *transfinite*. A set n is an aleph iff n is an infinite ordinal number and n is not the same size as any lesser ordinal. Since ω is an infinite ordinal number, and ω is not the same size as any finite ordinal, ω is an aleph. Hence $\omega = \aleph_0$.

We've defined many countable ordinals. These are ω, $\omega+n$, $\omega \cdot n$, and so on. Suppose X is the set of all countable ordinals. An ordinal is countable iff it is smaller than $\aleph_0$ or it is the same size as $\aleph_0$. Thus X = $\{ x \mid x \lesssim \aleph_0 \}$. Since X includes every finite ordinal, we know that X is either the same size as or bigger than $\aleph_0$. Is X the same size as $\aleph_0$? Can X be put into a 1-1 correspondence with $\aleph_0$? If so, then X is a member of X. But no set is a member of itself. So X can't be put into a 1-1 correspondence with $\aleph_0$. It follows that X is larger than $\aleph_0$. The number X is the next infinite cardinal. Just as 1 is the next cardinal after 0, so let us say that $\aleph_1$ is the next cardinal after $\aleph_0$. We define

$\aleph_1$ = the set of all ordinals smaller than $\aleph_0$ or the same size as $\aleph_0$.

Since any ordinal smaller than or the same size as $\aleph_0$ is countable, it follows that

$\aleph_1 = \{ x \mid x$ is a countable ordinal $\}$.

And since any ordinal that is smaller than or the same size as $\aleph_0$ is smaller than $\aleph_1$, the number $\aleph_1$ fits our definition of ordinal numbers (it fits the definition that the ordinal n is the set of all numbers less than n). That is,

$\aleph_1$ = the set of all numbers smaller than $\aleph_1$.

In other words, $\aleph_1$ is a number. It is an ordinal number. But since every ordinal that is smaller than $\aleph_1$ is a member of $\aleph_1$, $\aleph_1$ is the smallest ordinal number that is the same size as $\aleph_1$. Thus $\aleph_1$ is a cardinal number. $\aleph_1$ is the smallest cardinal that is greater than every countable ordinal (and thus greater than every countable cardinal). $\aleph_1$ is the least uncountable cardinal. Note that there cannot be any cardinal between $\aleph_0$ and $\aleph_1$. For if there were, it would be countable. And then its cardinality would be $\aleph_0$. We can repeat this reasoning to generate the series of transfinite cardinals:

$\aleph_2$ = the set of all numbers less than $\aleph_1$ or the same size as $\aleph_1$;

$\aleph_2$ = the set of all numbers less than $\aleph_2$.

As a rule, we say $\aleph(x)$ is the set of all ordinals less than x or the same size as x. In symbols, $\aleph(x) = \{ y \mid y \precsim x\}$. By the argument above, $\aleph(x)$ is bigger than x. If x is an infinite cardinal, then $\aleph(x)$ is the next infinite cardinal greater than x. We have:

$$\aleph_1 = \aleph(\aleph_0) = \text{the set of all numbers less than } \aleph_0 \text{ or the same size as } \aleph_0;$$

$$\aleph_2 = \aleph(\aleph_1) = \text{the set of all numbers less than } \aleph_1 \text{ or the same size as } \aleph_1;$$

$$\aleph_3 = \aleph(\aleph_2) = \text{the set of all numbers less than } \aleph_2 \text{ or the same size as } \aleph_2.$$

There is a series of alephs:

$$\aleph_0 \quad \aleph_1 \quad \aleph_2 \quad \aleph_3 \ldots \aleph_n \quad \aleph_{n+1} \ldots \aleph_\omega \ldots$$

There are infinitely many ordinals between any two alephs. It's just like fractions between whole numbers: the alephs are like the whole numbers. Of course, while this has all been fun, it's also been informal. But axiomatic set theories allow us to prove the existence of an endless series of alephs.

We've now defined two sequences of transfinite numbers: the alephs and the beths. We know from advanced set theory that every transfinite cardinal is an aleph. So every beth is an aleph. For example, there is some aleph $\aleph_n$ such that $\aleph = \aleph_n$. In other words, there is some $\aleph_n$ such that $\aleph_n = 2^\wedge\aleph_0$. However, the standard rules of set theory (the Zermelo-Fraenkel-Choice axioms) do not say anything definite about how to associate the alephs with the beths. Many possible associations are consistent with these axioms. Many mathematicians and philosophers regard this as an undesirable vagueness. Consequently, one of the outstanding problems in the theory of transfinite numbers, and one of the main outstanding problems of set theory generally, is to figure out how to precisely associate the beths with the alephs. This is the *continuum problem*.

7. Transfinite Recursion

7.1 Rules for the Long Line

We've used Cantor's number generating rules to define the ordinal number line. We've explicitly discussed the finite numbers and the least infinite number ω. But we've also defined a long line of numbers beyond ω. It's worth looking at Cantor's rules again, pointing out some of our recently defined transfinite numbers:

Initial Rule. There exists an initial ordinal 0.

Successor Rule. For every ordinal n, there exists a successor ordinal $n+1$. The successor rule generates all the positive finite ordinals 1, 2, 3, and so on. It also generates the transfinite successor ordinals. Here are a few examples of transfinite successors: $\omega+1$, $\omega+2$, $\omega+n$, $(\omega \cdot n)+1$, and so on.

Limit Rule. For any endless series of increasingly large ordinals, there exists a limit ordinal greater than every ordinal in that series. The least limit ordinal is ω. But there are greater limit ordinals. For example, every ordinal of the form $\omega \cdot n$ is a limit ordinal; every ordinal of the form ω^n is a limit ordinal. All the alephs are limit ordinals. So $\aleph_1, \aleph_2, \ldots \aleph_n, \ldots \aleph_\omega$ are all limit numbers. All the beth numbers are limit ordinals. The series of limit ordinals is endless in a *big* way.

The *maximal ordinal line* is that line of ordinal numbers than which no longer is logically possible. It includes the initial ordinal 0. It includes every finite ordinal. It includes the least limit ordinal ω. It includes every possible transfinite successor ordinal and every possible limit ordinal. It includes all the alephs and all the beths. For full precision, we need to use all the axioms of set theory to define the maximal ordinal line. Indeed, if you want to get really technical, the maximal ordinal line includes all the *large cardinals* (see Drake, 1974). These are ordinals (every cardinal is an ordinal) that are so big that they cannot be defined in terms of any set-theoretic operations on lesser ordinals. They cannot be reached from below. Each large cardinal has to be introduced with its own special set-theoretic axiom – much like ω. Go learn about large cardinals!

7.2 The Sequence of Universes

We've given recursive definitions that rise through all the finite ordinals and end with the least limit ordinal ω. But we've also defined a long line of ordinals beyond ω. Any ordinal on the maximal ordinal line is either a successor ordinal or a limit ordinal. We can extend the notion of a recursive definition into the transfinite by giving rules that hold at all successor and limit ordinals. A definition that associates every ordinal – whether finite or transfinite – with an object is a definition by *transfinite recursion.*

We illustrate transfinite recursion with possible universes. According to Leibniz, there is a single best possible universe (*Monadology*, 53-55). Against this idea, it is often said that there can't be any best of all possible universes. For any universe, you can define a better universe. Consequently, there is an endless series of increasingly better possible universes (Reichenbach, 1979; Fales, 1994). Let's use transfinite recursion to define an endless series of ever better universes. As expected, we're not concerned with whether or not these universes really exist; we're just illustrating a mathematical technique. The series of ever more perfect universes is defined by transfinite recursion like this:

> *Initial Rule.* For the initial ordinal 0 on the maximal ordinal line, there exists an initial universe U(0). The initial universe is the least perfect universe.
>
> *Successor Rule.* For every successor ordinal $n+1$ on the maximal ordinal line, there exists a successor universe U($n+1$). For any $n+1$, the successor universe U($n+1$) is an improvement of its predecessor universe U(n).
>
> *Limit Rule.* For every limit ordinal L on the maximal ordinal line, there exists a limit universe U(L). The universe U(L) is better than every universe in the series of which it is the limit. It is an improvement of the whole series. There are limit universes for all the alephs: U($\aleph_0$), U($\aleph_1$), . . . U($\aleph_n$), . . . U($\aleph_\omega$), and so on.

7.3 Degrees of Divine Perfection

Philosophy of religion can use transfinite recursion. God is said to be ontologically maximal. For any property P, if God has P, that is, if P is one of the divine perfections, then the degree to which God has P is maximal. And it must be maximal in the greatest possible sense. For any divine perfection P, the best way to define P is to use transfinite recursion to define a series of degrees of P. These degrees approach the absolutely maximal degree of P. It should be clear that we take no position on the existence of God. We're only interested in mathematical modeling. We're just extending the degrees of perfection examples from Chapter 2, section 10 and Chapter 7, section 3.4.

For every ordinal n, we define an n-th degree of divine perfection. But what is divine perfection? We can take an easy approach to this hard idea: divine perfection is divine creativity. Accordingly, we'll use transfinite recursion to define an endless series of degrees of divine creativity. Each degree of divine creativity is associated with some collection of created objects. What should these objects be? Since we've already used transfinite recursion to define an endless series of universes, you won't be surprised if we let them be universes. Let P be the perfection of divine creativity. We might say that P(God, n) includes only universe U(n). Hence each greater degree of divine creativity is associated with the creation of a better universe. But all universes have some perfection. So it seems better to associate the n-th degree of divine creativity with the set of all universes less than n. We now define P by transfinite recursion as follows:

> *Initial Rule.* For the initial ordinal 0, there exists an initial degree of the divine perfection P. This is P(God, 0). It's reasonable to think that the 0

degree of divine creativity is the null or empty degree. It does not include any universes. More precisely, P(God, 0) = {}.

Successor Rule. For every successor ordinal n+1 on the maximal ordinal line, there exists a successor degree of the perfection P. This is P(God, n+1). Just as the ordinal n+1 is defined in terms of n, so P(God, n+1) is defined in terms of P(God, n). Each next degree of P is an extension or amplification of the previous degree. To extend P(God, n) to P(God, n+1), we just add universe U(n). Thus P(God, n+1) = P(God, n) $\cup$ {U(n)}. For example,

P(God, 1) = P(God, 0) $\cup$ {U(0)} = {U(0)};
P(God, 2) = P(God, 1) $\cup$ {U(1)} = {U(0), U(1)};
P(God, 3) = P(God, 2) $\cup$ {U(2)} = {U(0), U(1), U(2)}.

As a rule, P(God, n+1) = {U(0), . . . U(n)}. And since n+1 = {0, . . . n}, it follows that for any successor ordinal n+1, P(God, n+1) = {U(i) | $i \in n$+1}.

Limit Rule. For every limit ordinal L on the maximal ordinal line, there exists a limit degree of the perfection P. This is P(God, L). Just as the ordinal L is defined in terms of all the ordinals less than L, so P(God, L) is defined in terms of P(God, i) for all i less than L. We let P(God, L) be the set of all universes whose indexes are less than L. Recall that x < L iff $x \in$ L. Hence P(God, L) = { U(i) | $i \in$ L}.

We've defined degrees of divine creativity for all ordinals on the maximal ordinal line. It's pretty easy to see that for any ordinal x,

P(God, x) = {U(i) | $i \in x$}.

The sequence of degrees indexed by ordinals rises to a maximal degree of the perfection P. It rises to a maximal degree of divine creativity. This degree is not indexed by any ordinal. It's just P(God). P(God) is defined in terms of P(God, k) for every ordinal k on the maximal ordinal line. Formally, P(God) = {U(i) | i is an ordinal}.

We can express this more extensionally by using the collection of all ordinals. This collection is denoted Ω. The collection of all ordinals is too general to be a set. To be sure, Ω is a proper class. The perfection P(God) has the rank Ω. It has the rank of a proper class. The proper class of ordinals is absolutely infinite. Hence P(God) is an absolutely infinite perfection. It looks like this:

P(God) = {U(i) | $i \in \Omega$ }.

Further Study

We've mentioned many opportunities for further study in the text. Here are a few other opportunities.

On the Web

Additional resources for *More Precisely* are available on the World Wide Web. These resources include extra examples as well as exercises. For more information, please visit

<http://broadviewpress.com/moreprecisely>

or

<http://www.ericsteinhart.com>.

Sets & Relations

Much of what we've covered in our discussion of sets and relations falls within the scope of *discrete mathematics*. Rosen (1999) is an excellent text in discrete math. Hamilton (1982) is a good introductory book on set theory, including class theory. Devlin (1991) is a good book at a more advanced level.

Machines

Grim et al. (1998) contains many examples of the use of computers – and thus finite state machines – in many different branches of philosophy, including logic and ethics. From finite machines, it's natural to move to Turing Machines. Boolos & Jeffrey (1989) is an excellent and extensive discussion of Turing Machines.

Semantics

Chierchia & McConnell-Ginet (1991) provide a good textbook on formal semantics. Their text aims at modeling real English. Many philosophers have long noticed the parallels between modality and temporality. For example, something is impossible iff it *never* happens; it is necessary iff it *always* happens; it is contingent iff it *sometimes* happens and *sometimes* does not happen; it is possible iff it sometimes happens. Times act like worlds. Sider (1996, 2001) has worked out a temporal version of counterpart theory.

Utilitarianism

Feldman (1997) provides the basis for most of our Utilitarianism chapter. Broome (1991) is a more advanced text on mathematical utilitarianism.

Probability

Skyrms (1966) is old but it's a worthy classic. More recently, Hacking (2001) is an outstanding introduction to probability and inductive reasoning. It is well-written, filled with examples, and carefully examines the relevant philosophical issues. Mellor (2005) discusses the philosophical interpretations of probability. Howson & Urbach (2005) discuss the uses of Bayesianism in scientific reasoning.

Infinity

Aczel (2000) is an excellent historical introduction to the modern theory of infinity. It's fun, readable, and covers lots of mathematical and philosophical ground. Barrow (2005) is an accessible review of the uses of infinity in mathematics and physics.

Glossary of Symbols

Symbol	Meaning	Page
$x \in y$	x is a member of y	2
$x \notin y$	x is not a member of y	2
$x = y$	x is identical with y	3
iff	if and only if	3
$\{ x \mid x \text{ is P}\}$	the set of all x such that x is P	4
$X \subseteq Y$	X is a subset of Y	4
$X \subset Y$	X is a proper subset of Y	5
$X \supseteq Y$	X is a superset of Y	6
$\{x\}$	the unit set of x	6
$\{\}$	the empty set	6
$\varnothing$	the empty set	6
$X \cup Y$	the union of X and Y	8
$X \cap Y$	the intersection of X and Y	9
$X - Y$	the difference between X and Y	10
$\cup X$	the union of all the sets in X	12
$\mathcal{P}X$	the power set of X	13
pow X	the power set of X	13
V_k	the k-th rank in the iterative hierarchy V	17
V	the proper class of all sets	17
(x, y)	the ordered pair of x and y	21
$X \times Y$	the Cartesian product of X and Y	23
R^{-1}	the inverse of the relation R	26

$[x]$	the equivalence class of x	28
$R \circ R$	the composition of R with itself	30
R^n	the n-th power of the relation R	31
R^*	the transitive closure of R	31
$x << y$	x is a part of y	48
$f: X \to Y$	the function from X to Y	49
$f(x)$	the value of the application of f to x	49
f^{-1}	the inverse of function f	52
$\sum_{x \in A} x$	the sum, for all x in A, of x	59
$\{S_n\}$	the sequence S_0 to S_n	60
$\sum_{i=0}^{n} S_i$	the sum, for i varying from 0 to n, of S_i	60
$\bigcup_{i=0}^{n} S_i$	the union, for i varying from 0 to n, of S_i	61
$\bigcap_{i=0}^{n} S_i$	the intersection, for i varying from 0 to n, of S_i	61
$\|S\|$	the cardinality of S	61
#S	the cardinality of S	61
P(E)	the probability of event E	110
P(E | H)	the probability of E given H	117
Ex(A)	the expected utility of act A	135
ω	the least infinite number (same as $\aleph_0$)	151
$X \precsim Y$	X is smaller than or as big as Y	164
$\aleph_0$	the least infinite number (same as ω)	166

$\aleph$	the cardinality of pow ω	175
$\aleph_1$	the least uncountable number	177
Ω	the proper class of all numbers	181

References

Aczel, P. (1988) *Non-Well-Founded Sets*. CSLI Lecture Notes 14. Stanford, CA: CSLI Publications.

Aczel, A. (2000) *The Mystery of the Aleph*. New York: Washington Square Press.

Anselm (1076) *Monologion*. In B. Davies & G. Evans (Eds.) (1998) *Anselm of Canterbury: The Major Works*. New York: Oxford University Press.

Augustine (1993) *On the Free Choice of the Will*. Trans. T. Williams. Indianapolis: Hackett.

Barrow, J. (2005) *The Infinite Book*. New York: Pantheon Books.

Benacerraf, P. (1962) Tasks, super-tasks, and the modern Eleatics. *Journal of Philosophy 59* (24), 765-84.

Benacerraf, P. (1965) What numbers could not be. In P. Benacerraf & H. Putnam (Eds.) (1984) *Philosophy of Mathematics*. New York: Cambridge University Press, 272-95.

Black, M. (1952) The identity of indiscernibles. In J. Kim & E. Sosa (Eds.) (1999) *Metaphysics: An Anthology*. Malden, MA: Blackwell, 66-71.

Boolos, G. (1971) The iterative conception of set. In P. Benacerraf & H. Putnam (Eds.) (1996) *Philosophy of Mathematics: Selected Readings*. Second Edition. New York: Cambridge University Press, 486-502.

Boolos, G. & Jeffrey, R. (1989) *Computability and Logic*. Third Edition. New York: Cambridge University Press.

Bostrom, N. (2003) Are you living in a computer simulation? *Philosophical Quarterly 53* (211), 243-55.

Boyd, R. (1984) The current status of scientific realism. In J. Leplin (Ed.) (1984), *Scientific Realism*. Berkeley, CA: University of California Press, 41-82.

Broome, J. (1991) *Weighing Goods: Equality, Uncertainty, and Time*. Cambridge, MA: Basil Blackwell.

Burks, A. (1948-49) Icon, index, and symbol. *Philosophy and Phenomenological Research 9*, 673-89.

Burks, A. (1973) Logic, computers, and men. *Proceedings and Addresses of the American Philosophical Association 46,* 39-57.

Cantor, G. (1955) *Contributions to the Founding of the Theory of Transfinite Numbers.* Trans. P. Jourdain. New York: Dover.

Casati, R. & Varzi, A. (1999) *Parts and Places: The Structures of Spatial Representation.* Cambridge, MA: MIT Press.

Chierchia, G. & McConnell-Ginet, S. (1991) *Meaning and Grammar: An Introduction to Semantics.* Cambridge, MA: MIT Press.

Copeland, B.J. (1998a) Super-Turing machines. *Complexity 4* (1), 30-32.

Copeland, B.J. (1998b) Even Turing machines can compute uncomputable functions. In C. Calude, J. Casti, and M. Dinneen (Eds.) (1998), *Unconventional Models of Computation.* New York: Springer-Verlag.

Copeland, B.J. (2000) Narrow versus wide mechanism. *Journal of Philosophy 95* (1), 5-32.

Dennett, D. (1991) Real patterns. *Journal of Philosophy 88* (1), 27-51.

Devlin, K. (1991) *The Joy of Sets.* New York: Springer-Verlag.

Drake, F. (1974) *Set Theory: An Introduction to Large Cardinals.* New York: American Elsevier.

Dretske, F. (1981) *Knowledge and the Flow of Information.* Cambridge, MA: MIT Press.

Fales, E. (1994) Divine freedom and the choice of a world. *International Journal for the Philosophy of Religion 35,* 65-88.

Feldman, F. (1997) *Utilitarianism, Hedonism, and Desert.* New York: Cambridge University Press.

Fredkin, E. (1991) Digital mechanics. In Gutowitz, H. (Ed.) (1991) *Cellular Automata: Theory and Experiment.* Cambridge, MA: MIT Press, 254-70.

Fredkin, E. (2003) An introduction to digital philosophy. *International Journal of Theoretical Physics 42* (2), 189-247.

Frege, G. (1884) The concept of number. In P. Benacerraf & H. Putnam (1996), *Philosophy of Mathematics.* Second Edition. New York: Cambridge University Press, 130-59.

Goodman, N. (1956) A world of individuals. In I.M. Bochenski, A. Church, N. Goodman (Eds.), *The Problem of Universals*. Notre Dame: University of Notre Dame Press, 13-32.

Grim, P., Mar, G., & St. Denis, P. (1998) *The Philosophical Computer: Exploratory Essays in Philosophical Computer Modeling*. Cambridge, MA: MIT Press.

Hacking, I. (2001) *An Introduction to Probability and Inductive Logic*. New York: Cambridge University Press.

Hallett, M. (1988) *Cantorian Set Theory and Limitation of Size*. New York: Oxford University Press.

Hamilton, A. (1982) *Numbers, Sets, and Axioms: The Apparatus of Mathematics*. New York: Cambridge University Press.

Hobbes, T. (1651/1962). *Leviathan: or The Matter, Forme, and Power of a Common Wealth Ecclesiasticall and Civil*. M. Oakeshott (Ed.). New York: Oxford University Press.

Howson, C. & Urbach, P. (2005) *Scientific Reasoning: The Bayesian Approach*. Third Edition. La Salle, IL: Open Court Press.

Hume, D. (1990) *A Treatise of Human Nature*. Second Edition. New York: Oxford University Press.

Kim, J. (1998). *Mind in a Physical World: An Essay on the Mind-Body Problem and Mental Causation*. Cambridge, MA: MIT Press.

Kirk, G. & Raven, J. (1957) *The Presocratic Philosophers*. New York: Cambridge University Press.

Koetsier, T. & Allis, V. (1997) Assaying supertasks. *Logique et Analyse 159*, 291-313.

La Mettrie, J.O. de (1748 / 1999) *Man a Machine*. Chicago: Open Court Publishing.

Lewis, D. (1968) Counterpart theory and quantified modal logic. *Journal of Philosophy 65*, 113-26.

Lewis, D. (1976) Survival and identity. In A.O. Rorty (Ed.), *The Identities of Persons*. Berekeley, CA: University of California Press, 17-40.

Lewis, D. (1991) *Parts of Classes*. Cambridge, MA: Blackwell.

Locke, J. (1690/1959) *An Essay Concerning Human Understanding*. New York: Dover Publications.

Loux, M. (2002) *Metaphysics: A Contemporary Introduction*. New York: Routledge.

Lovejoy, A. (1936) *The Great Chain of Being*. Cambridge, MA: Harvard University Press.

Mellor, D. (2005) *Probability: A Philosophical Introduction*. New York: Routledge.

Parfit, D. (1971) Personal identity. *Philosophical Review 80,* 3-27.

Perry, J. (1976) The problem of personal identity. In J. Perry (Ed.) *Personal Identity*. Berkeley, CA: University of California Press, 3-30.

Poundstone, W. (1985) *The Recursive Universe: Cosmic Complexity and the Limits of Scientific Knowledge*. Chicago: Contemporary Books, Inc.

Quine, W.V.O. (1976) Wither physical objects? *Boston Studies in the Philosophy of Science 39,* 497-504.

Quine, W.V.O. (1978) Facts of the matter. *Southwestern Journal of Philosophy 9* (2), 155-69.

Quine, W.V.O. (1981) *Theories and Things*. Cambridge, MA: Harvard University Press.

Quine, W.V.O. (1986) Reply to Charles D. Parsons. In L. Hahn & P. Schilpp (Eds.) *The Philosophy of W.V. Quine*. La Salle, IL: Open Court, 396-404.

Reichenbach, B. (1979) Must God create the best possible world? *International Philosophical Quarterly 19* (2), 203-12.

Reid, T. (1975) Of Mr. Locke's account of our personal identity. In J. Perry (Ed.) *Personal Identity*, Berkeley CA: University of California Press, 107-12.

Rendell, P. (2002) Turing universality of the game of life. In A. Adamatzky (Ed.) (2002), *Collision-based Computation*. New York: Springer, 513-39.

Rescher, N. (1991) *G.W. Leibniz's Monadology: An Edition for Students*. Pittsburgh, PA: University of Pittsburgh Press.

Resnik, M. (1997) *Mathematics as a Science of Patterns*. New York: Oxford University Press.

Rosen, K. (1999) *Discrete Mathematics and its Applications.* Fourth edition. Boston: McGraw-Hill.

Royce, J. (1927) *The World and the Individual, First Series, Supplementary Essay.* New York: The Macmillan Company.

Shapiro, S. (1997) *Philosophy of Mathematics: Structure and Ontology.* New York: Oxford University Press.

Sider, T. (1996) All the world's a stage. *Australasian Journal of Philosophy 74,* 433-53.

Sider, T. (2001) *Four-Dimensionalism: An Ontology of Persistence and Time.* New York: Oxford University Press.

Skyrms, B. (1966) *Choice and Chance: An Introduction to Inductive Logic.* Belmont, CA: Dickenson Publishing.

Steinhart, E. (2003) Why numbers are sets. *Synthese 133,* 343-61.

Thomson, J. (1954) Tasks and supertasks. *Analysis 15,* 1-13.

Toffoli, T. & Margolus, N. (1987) *Cellular Automata Machines: A New Environment for Modeling.* Cambridge, MA: MIT Press.

Turing, A. (1936) On computable numbers, with an application to the Entscheidungsproblem. *Proceedings of the London Mathematical Society 2* (42), 230-65.

van Heijenoort, J. (Ed.) (1967) *From Frege to Godel: A Source Book in Mathematical Logic, 1879-1931.* Cambridge, MA: Harvard University Press.

von Neumann, J. (1923) On the introduction of transfinite numbers. In J. van Heijenoort (Ed.) (1967), 346-54.

Wang, H. (1974) The concept of set. In P. Benacerraf & H. Putnam (Eds.) (1996) *Philosophy of Mathematics: Selected Readings.* Second Edition. New York: Cambridge University Press, 530-70.

Wiggins, D. (1976) Locke, Butler and the stream of consciousness: And men as a natural kind. In A.O. Rorty (Ed.), *The Identities of Persons.* Berekeley, CA: University of California Press, 139-73.

Index